T0195518

Analyzing Contingency Contracting Purchases for Operation Iraqi Freedom
(Unrestricted Version)

Laura H. Baldwin • *John A. Ausink* • *Nancy F. Campbell*
John G. Drew • *Charles Robert Roll, Jr.*

Prepared for the United States Air Force

Approved for public release; distribution unlimited

PROJECT AIR FORCE

The research described in this report was sponsored by the United States Air Force under Contract FA7014-06-C-0001. Further information may be obtained from the Strategic Planning Division, Directorate of Plans, Hq USAF.

Library of Congress Cataloging-in-Publication Data

Analyzing contingency contracting purchases for operation Iraqi Freedom (unrestricted version) / Laura H. Baldwin ... [et al.].
    p. cm.
    Includes bibliographical references.
    ISBN 978-0-8330-4234-7 (pbk. : alk. paper)
    1. Defense contracts—United States. 2. United States. Air Force—Procurement.
3. United States. Air Force—Costs. 4. Iraq War, 2003– I. Baldwin, Laura H., 1967–

UG633.2.A63 2007
956.7044'348—dc22

                                         2007041603

Published 2008 by the RAND Corporation
1776 Main Street, P.O. Box 2138, Santa Monica, CA 90407-2138
1200 South Hayes Street, Arlington, VA 22202-5050
4570 Fifth Avenue, Suite 600, Pittsburgh, PA 15213-2665
RAND URL: http://www.rand.org/
To order RAND documents or to obtain additional information, contact
Distribution Services: Telephone: (310) 451-7002;
Fax: (310) 451-6915; Email: order@rand.org

Preface

This monograph documents RAND Corporation research on purchases made to support U.S. Air Force activities during Operation Iraqi Freedom (OIF). We provide a baseline of contingency contracting activities for OIF and insights relevant to important contracting and logistics policy issues. These analyses are based primarily on data from the U.S. Central Command Air Forces (CENTAF) comptroller describing purchases made during FYs 2003 and 2004 at CENTAF locations supporting OIF.

This research was part of a broader study titled "Contracting to Support Contingencies: Lessons from Recent Operations," sponsored by the Air Force Deputy Assistant Secretary for Contracting (SAF/AQC) and the Deputy Chiefs of Staff for Logistics, Installations, and Mission Support, Resource Integration (AF/A4/7P) and Logistics Readiness (AF/A4R). The study was conducted within the Resource Management Program of RAND Project AIR FORCE. An expanded version of this document, available to U.S. Department of Defense personnel and contractors, contains additional information and data (see Baldwin, Ausink, Campbell, et al., 2008).

This monograph is designed to assist contracting and logistics policymakers in their efforts to improve future Air Force contingency contracting activities. RAND Project AIR FORCE has previously examined issues related to Air Force contracting and logistics policy. The resulting publications in the area of agile combat support include the following:

- *A Framework for Enhancing Airlift Planning and Execution Capabilities Within the Joint Expeditionary Movement System*, Robert S. Tripp, Kristin F. Lynch, Charles Robert Roll, Jr., John G. Drew, and Patrick Mills (MG-377-AF).
- *Strategic Analysis of Air National Guard Combat Support and Reachback Functions*, Robert S. Tripp, Kristin F. Lynch, Ronald G. McGarvey, Don Snyder, Raymond A. Pyles, William A. Williams, and Charles Robert Roll, Jr. (MG-375-AF).
- *Unmanned Aerial Vehicle End-to-End Support Considerations*, John G. Drew, Russell D. Shaver, Kristin F. Lynch, Mahyar A. Amouzegar, and Don Snyder (MG-350-AF).
- *Supporting Air and Space Expeditionary Forces: Lessons from Operation Iraqi Freedom*, Kristin F. Lynch, John G. Drew, Robert S. Tripp, and Charles Robert Roll, Jr. (MG-193-AF).
- *Supporting Air and Space Expeditionary Forces: Analysis of Combat Support Basing Options*, Mahyar A. Amouzegar, Robert S. Tripp, Ronald G. McGarvey, Edward W. Chan, and Charles Robert Roll, Jr. (MG-261-AF).
- *Supporting Air and Space Expeditionary Forces: Lessons from Operation Enduring Freedom*, Robert S. Tripp, Kristin F. Lynch, John G. Drew, and Edward W. Chan (MR-1819-AF).
- *Supporting Air and Space Expeditionary Forces: A Methodology for Determining Air Force Deployment Requirements*, Don Snyder and Patrick Mills (MG-176-AF).
- *Supporting Expeditionary Aerospace Forces: An Operational Architecture for Combat Support Execution Planning and Control*, James Leftwich, Robert Tripp, Amanda Geller, Patrick Mills, Tom LaTourrette, C. Robert Roll, Jr., Cauley von Hoffman, and David Johansen (MR-1536-AF).
- *Supporting Expeditionary Aerospace Forces: A Concept for Evolving the Agile Combat Support/Mobility System of the Future*, Robert S. Tripp, Lionel A. Galway, Timothy Ramey, Mahyar A. Amouzegar, and Eric Peltz (MR-1179-AF).
- *Supporting Expeditionary Aerospace Forces: New Agile Combat Support Postures*, Lionel A. Galway, Robert S. Tripp, Timothy Ramey, and John G. Drew (MR-1075-AF).

- *Supporting Expeditionary Aerospace Forces: An Integrated Strategic Agile Combat Support Planning Framework*, Robert S. Tripp, Lionel A. Galway, Paul Killingsworth, Eric Peltz, Timothy Ramey, and John G. Drew (MR-1056-AF).

Recent RAND Project AIR FORCE work in the area of purchasing and supply management includes the following:

- *An Assessment of Air Force Data on Contract Expenditures*, Lloyd Dixon, Chad Shirley, Laura H. Baldwin, John A. Ausink, and Nancy F. Campbell (MG-274-AF).
- *Air Force Service Procurement: Approaches for Measurement and Management*, Laura H. Baldwin, John A. Ausink, and Nancy Nicosia (MG-299-AF).
- *Air Force Procurement Workforce Transformation: Lessons from the Commercial Sector*, John A. Ausink, Laura H. Baldwin, and Christopher Paul (MG-214-AF).
- *Using a Spend Analysis to Help Identify Prospective Air Force Purchasing and Supply Management Initiatives: Summary of Selected Findings*, Nancy Y. Moore, Cynthia R. Cook, Clifford A. Grammich, and Charles Lindenblatt (DB-434-AF).
- *Implementing Best Purchasing and Supply Management Practices: Lessons from Innovative Commercial Firms*, Nancy Y. Moore, Laura H. Baldwin, Frank Camm, and Cynthia R. Cook (DB-334-AF).
- *Implementing Performance-Based Services Acquisition (PBSA): Perspectives from an Air Logistics Center and a Product Center*, John A. Ausink, Laura H. Baldwin, Sarah Hunter, and Chad Shirley (DB-388-AF).
- *Performance-Based Contracting in the Air Force: A Report on Experiences in the Field*, John A. Ausink, Frank Camm, and Charles Cannon (DB-342-AF).
- *Strategic Sourcing: Measuring and Managing Performance*, Laura H. Baldwin, Frank Camm, and Nancy Y. Moore (DB-287-AF).

- *Incentives to Undertake Sourcing Studies in the Air Force*, Laura H. Baldwin, Frank Camm, Edward G. Keating, and Ellen M. Pint (DB-240-AF).
- *Strategic Sourcing: Theory and Evidence from Economics and Business Management*, Ellen M. Pint and Laura H. Baldwin (MR-865-AF).

RAND Project AIR FORCE

RAND Project AIR FORCE, a division of the RAND Corporation, is the U.S. Air Force's federally funded research and development center for studies and analyses. RAND Project AIR FORCE provides the Air Force with independent analyses of policy alternatives affecting the development, employment, combat readiness, and support of current and future aerospace forces. Research is conducted in four programs: Aerospace Force Development; Manpower, Personnel, and Training; Resource Management; and Strategy and Doctrine.

Additional information about RAND Project AIR FORCE is available on our Web site: http://www.rand.org/paf/

Contents

Figures

Tables

Summary

Contingency contracting officers (CCOs) serve a vital role in contingency operations: They are given the authority to enter into, administer, and terminate contracts on behalf of the government in support of contingency operations. They also act as business advisors to the deployed on-scene commander.

Since September 11, 2001, the U.S. Air Force has been involved in two significant contingency operations in the U.S. Central Command (CENTCOM) area of responsibility (AOR): Operation Enduring Freedom (OEF) in Afghanistan and OIF in Iraq. After early experiences in both OEF and OIF, SAF/AQC asked RAND Project AIR FORCE to gather and analyze data on goods and services purchased to support Air Force missions in OIF in an effort to determine (1) the size and extent of contractor support and (2) how plans for and the organization and execution of contingency contracting activities might be improved so that CCOs can better support the warfighter in future operations.

The motivation for undertaking this study was twofold. First, the contracting community did not have a comprehensive, detailed database of contingency purchases that would allow analyses of the types and amounts of goods and services purchased to support Air Force mission activities. Second, it was thought that insights from analyses of recent contingency contracting experiences would help inform decisions about a number of important policy issues related to planning, training, and CCO assignments. Such data could also be used to seek improvements in purchasing practices across the theater.

Contingency purchases associated with OIF were made by a large number of organizations around the world. The analyses presented here are based on CCO purchases occurring at purchasing organizations located within the CENTCOM AOR that supported OIF during FYs 2003 and 2004. Data on these purchases were obtained from transaction logs maintained by the office of the CENTAF comptroller, headquartered at Shaw AFB, South Carolina.[1] These data include more than 24,000 transactions obligating more than $300 million.

In this monograph, we provide a baseline analysis of purchases (pp. 17–44). We describe the details of these transactions in terms of

- *who* (which organizations) made purchases
- *what* types of goods and services were purchased
- *when* the purchases were made (time periods)
- *how* the purchases were made (contracting tools used)
- *from whom* (suppliers) the purchases were made.

We then use these data to illustrate how such analyses can be applied to improve (1) the alignment of contracting personnel with demands in theater (pp. 45–48), (2) the preparation of CCOs prior to deployment so that they can more efficiently and effectively satisfy the requirements of combat forces (pp. 48–50), (3) the ability of combat support planners to make trade-offs between advanced purchasing and management of resources and purchasing in theater as needed (pp. 50–53), and (4) the sharing of lessons among CCOs in theater (pp. 53–55).

Selected Findings from Data Analyses

We examined CCO expenditures for each of the 24 purchasing organizations supporting OIF during FYs 2003 and 2004 for which we had data (pp. 19–21). Expenditures at Al Udeid AB, Qatar, for both the air wing and the Combined Air Operations Center (CAOC) far exceeded

---

[1]   See Appendix B for a list of other spending sources for OIF activities.

those of other organizations, with the air wing spending approximately five times the amount spent by Al Dhafra AB, United Arab Emirates, the next-highest-spending organization. Normalizing for the number of months each purchasing organization was active during this period, we found that Al Udeid's air wing still had the largest expenditures, with monthly expenditures approximately 2.5 times those of the next-highest-spending organizations, the CAOC and Tabuk AB (pp. 19–21).

We grouped purchases according to 45 categories of goods and services. The top categories in terms of expenditures were construction supplies, vehicles, and construction services. We found that purchasing organizations spent, on average, more than 10 percent of their expenditures on construction supplies and vehicles and just under 10 percent on construction services (pp. 20–27).

The data indicate changing patterns of expenditures over time (pp. 27–33). In aggregate, spending was higher in FY 2003 than in FY 2004. The levels and time flow of expenditures varied between locations identified by CENTAF personnel as being "temporary" operating locations versus those thought to be more "permanent." A comparison of categories of spending at Baghdad International Airport and Tallil AB, two Iraqi bases with similar levels of expenditures, illustrates that the mix of goods and services purchased can be very different at seemingly similar bases. Information about the activities supported by the expenditures and the location itself would be needed to draw insights from the observed patterns.

Our analysis of types of payment vehicles utilized for transactions indicates that government purchase card (GPC) purchases represented more than one-third of the transactions made in FYs 2003 and 2004 but less than one-tenth of the dollars spent (pp. 33–37). GPCs were used for smaller transactions, primarily for goods rather than services. Blanket purchase agreements (BPAs) were used extensively by OIF purchasing organizations (pp. 38–41). They represented 30 percent of expenditures overall and more than 60 percent of contract expenditures for vehicles, heavy nonconstruction equipment, and food services.

The top 10 suppliers in terms of dollars represented more than 40 percent of the expenditures (pp. 41–44). At least two of these firms were identified by CENTAF personnel as "10-percenters," i.e., firms that mediate purchases for a fee. Based on firm names, many of the top suppliers for contract purchases appear to be regional firms, whereas GPCs were often used for purchases from Western firms. Top suppliers provided a wide range of goods and services to multiple purchasing organizations.

These types of descriptive analyses of expenditures can be used to motivate additional analyses utilizing supplemental data, such as descriptions of the physical locations of the purchasing activity (e.g., base population, extent and condition of infrastructure) and the mission activities supported by those purchases (e.g., numbers and types of aircraft).

We utilized data on numbers of CCOs deployed to selected purchasing organizations to examine expenditures per CCO and transactions per CCO (pp. 45–48). According to these simple workload measures, CCOs at Balad AB obligated more dollars and performed more transactions than did their counterparts at other Iraqi bases. However, obligations of CCOs at the CAOC were significantly higher. To interpret these findings, additional details are needed about how the nature of the work differs across locations.

Finally, we conducted a case study of water purchases to illustrate how detailed data on transactions can (1) enhance the ability of combat support planners to decide whether to make arrangements for purchasing selected resources in advance or to purchase as needed in theater and (2) provide helpful information to CCOs who are negotiating contracts across the AOR (pp. 50–55).

Our analyses demonstrate that a contingency contracting database such as that developed for this study can be a powerful and useful analytic tool. However, current practices concerning the documentation of CCO transactions in theater make the development of such information time-consuming. *We recommend that the Air Force (and the Department of Defense more broadly) establish a standardized methodology for collecting contingency contracting data on an ongoing basis to facilitate planning and policy decisions,* e.g., those associated with CCO staffing and training, combat support planning, and sharing of lessons within the theater for future contingencies. Such a data capability should incorporate detailed descriptions of individual transactions that can be easily sorted and aggregated for analyses (pp. 57–61). Table S.1 presents our recommendations for types of data to be collected.

Table S.1
Recommended Data to Be Collected on an Ongoing Basis

Type of Data	Explanation
Individual transactions	Data to be entered by purchasing CCO
Purchasing organization	Organization that purchases the goods or services
CCO	Individual responsible for the transaction
Recipient	Organization or location that benefited from the purchase, if different from the purchasing organization, e.g., base that benefited from a Rapid Engineer Deployable Heavy Operational Repair Squadron Engineers (RED HORSE) repair project
Text description	Description of full range of goods and services purchased through the transaction
Units	Number of goods purchased or period of time in which service is to be provided; break out according to types of goods or services covered by the transaction
Purchase category	General class(es) of goods or services purchased; break out according to types of goods or services covered by the transaction
Price	Price paid for the goods and services; when multiple goods and services are purchased within a single transaction, prices should be broken out by type
Supplier	Firm that provides the goods and services
Location of supplier	Identifies whether supplier is a local firm, regional firm, or other
Transaction ID	Unique identifier for the transaction, e.g., contract number
Payment mechanism	GPC or contract
Type of contract	For contracts, type of contract, e.g., BPA, Form SF44
Date of request	Date on which purchasing organization received the formal request for goods or services

Table S.1—Continued

Type of Data	Explanation
Date of payment	Date on which supplier was paid
Date of delivery	Date on which goods were delivered or services began
Comments	Any explanatory comments CCO deems useful
Activities supported by purchasing organizations	Supplemental data needed to explain purchasing trends; will vary over time
Population	Number of personnel supported by the purchasing organization
Mission activity	Description of mission activity supported by the purchasing organization's transactions, e.g., number and types of aircraft, special operations
Responsibility for base operating support (BOS)	Service branch responsible for providing BOS for the location
Infrastructure	Number of buildings or acres supported by the purchasing organization
Condition of infrastructure	Condition of infrastructure supported by the purchasing organization, particularly for new locations
Outlook	Plans for the purchasing organization, e.g., temporary operating location
Supply base	Supplemental data to facilitate improved purchasing over time
Supplier ratings	Performance ratings of suppliers (perhaps only key suppliers) based on, for example, the quality of goods and services, reliability, ease of working relationship

Acknowledgments

We thank Charlie E. Williams, Jr., formerly of SAF/AQC, for sponsoring this study. We are grateful to our main points of contact within the Contracting Operations Division (SAF/AQCK), Col David Glowacki, Col William David McKinney, Maj Robert Widmann, and Maj Randy Culbreth, for their guidance and assistance throughout this project.

In addition, we thank the many Air Force and Air Force contractor personnel who took time to teach us about contracting in a contingency environment and helped us understand the available data. This research would not have been possible without their help. Affiliations listed are those effective at the time of our research.

Assisting in this effort were Air Combat Command (ACC) staff Col Glenn Whitaker, Lt Col Robert Michael, Raymond Carpenter, and CMSgt Kevin Fraher; CENTAF staff Jim Evans, Lt Col Robert Blair, Lt Col Tonish Jones, Capt Michelle Griffith, Janetta S. Brown, Maj Chris Barker, Capt Doug Thrailkill, TSgt Ron Alexander, Maj Troy Sanders, Maj Mike Vaughn, Maj Scott Tomlinson, Maj Kristian Ellingsen, Charles S. Clark, Deanna Price, Joe Waldron, Bill Strickland, MSgt Marilyn Blanton, Moretta Cooper, Gerard Jones, Ken Trout, MSgt Dalton Tisdale, Glenda Faulknham, Steve Allen, Capt William Pendleton, and Maj Carlos Camarillo; Col John E. Cannaday of CENTCOM; Lt Col Kim Triesler of Air Force Logistics Management Agency (AFLMA); Air Force Civil Engineer Support Agency staff Jim Garred, Larry Edwards, TSgt Ken Longstreet, and Col Jo Worrell; U.S. Air Forces in Europe (USAFE) staff Col Mary Kringer, Maj Jen Block, MSgt Jeff Adkins, and MSgt Markus T. Gaines; Air Force

Special Operations Command (AFSOC) staff SMSgt Johnnie Jackson, MSgt Ferdinand Rodriguez, Bill Rone, Maj Michael Crook, Capt Trevor Tullie, Edward Shapiro, and CMSgt Tim Hulme; Air Mobility Command (AMC) staff CMSgt Ed Claunch, TSgt Tom Smith, and Dale Huegen; MSgt Ronald Godsy of Pacific Air Forces; SAF/AQC staff Col Maureen M. Clay, Col Wilma Slade, Lt Col Brett McMullen, and Lt Col Richard Unis; CMSgt Mike Doris of the Office of the Deputy Chief of Staff for Logistics, Installations, and Mission Support (AF/A4/7); John Johnson and Capt Joshua Weed of the Air Force Financial Management and Budget Office (SAF/FMBO); Thomas Kehoe of DynCorp; T. B. (Mac) McClelland of Center House, Ltd.; and Dwight Clark of RMS, LC (Johnson Controls).

We are also grateful for the support of our RAND colleagues. Christopher Paul provided valuable assistance in formatting the FY 2003 GPC data and conducting an initial literature review that helped shape Appendix A of this monograph. Lloyd Dixon helped collect data from AMC. Dahlia Lichter provided helpful research support. In addition, we thank Mahyar A. Amouzegar, Frank Camm, Louis Luangkesorn, Kristin F. Lynch, James Masters, Ronald G. McGarvey, Patrick Mills, Raymond A. Pyles, Don Snyder, and Robert S. Tripp for helping us understand the interactions between contingency contracting and agile combat support. Nancy Nicosia and Don Snyder provided helpful critiques of an earlier draft of this monograph. Finally, we thank Mary Debold for her skillful document assistance.

Abbreviations

AAFES	Army and Air Force Exchange Service
ACC	Air Combat Command
ACTT	Automated Contract Tracking Tool
AF/A4/7	Deputy Chief of Staff for Logistics, Installations, and Mission Support
AF/A4/7P	Deputy Chief of Staff for Logistics, Installations, and Mission Support, Resource Integration
AF/A4R	Deputy Chief of Staff for Logistics, Installations, and Mission Support, Logistics Readiness
AFAA	Air Force Audit Agency
AFCAP	Air Force Contract Augmentation Program
AFFARS	Air Force Federal Acquisition Regulation Supplement
AFI	Air Force Instruction
AFLMA	Air Force Logistics Management Agency
AFSOC	Air Force Special Operations Command
AMC	Air Mobility Command
AOR	area of responsibility

BEAR	Basic Expeditionary Airfield Resources
BOS	base operating support
BPA	blanket purchase agreement
CAOC	Combined Air Operations Center
CCO	contingency contracting officer
CENTAF	U.S. Central Command Air Forces
CENTCOM	U.S. Central Command
CONUS	continental United States
CORE	Cost-Oriented Resource Evaluator
CPS	Central Processing Site
DAU	Defense Acquisition University
DFAS	Defense Finance and Accounting Service
DoD	U.S. Department of Defense
EEIC	element of expense investment code
ESP	emergency and special program
GPC	government purchase card
GWOT	global war on terrorism
IMPAC	International Merchant Purchase Authorization Card
JCC	Joint Contract Center
LOGCAP	Logistics Civil Augmentation Program
MWR	morale, welfare, and recreation
NAICS	North American Industry Classification System
O&M	operations and maintenance
OEF	Operation Enduring Freedom

OIF	Operation Iraqi Freedom
ONW	Operation Northern Watch
OSW	Operation Southern Watch
PBSA	performance-based services acquisition
PIIN	procurement instrument identification number
PSAB	Prince Sultan Air Base
RED HORSE	Rapid Engineer Deployable Heavy Operational Repair Squadron Engineers
SAF/AQC	Air Force Deputy Assistant Secretary for Contracting
SAF/AQCK	Air Force Deputy Assistant Secretary for Contracting, Contracting Operations Division
SAF/FMBO	Air Force Financial Management and Budget Office
SIMM	single in-line memory module
SIPRNET	Secret Internet Protocol Router Network
TALCE	tanker airlift control element
UAV	unmanned aerial vehicle
USAFE	U.S. Air Forces in Europe
WRM	war reserve materiel

CHAPTER ONE

Introduction

Contractors have been an important part of U.S. war efforts since they were hired to take care of cavalry horses for the Continental Army during the Revolutionary War (Cahlink, 2003). While the history of contracted support to U.S. military operations is a long one, the role of that support has expanded rapidly and extensively, particularly since the end of the Cold War (Cahlink, 2003; Camm and Greenfield, 2005; CBO, 2005). Today, the U.S. Air Force, as well as the other U.S. military services, buys an enormous amount and variety of goods and services to support its contingency operations. These purchases are necessary for a wide range of activities, including feeding, housing, and protecting military personnel; repairing aircraft weapon systems; and transporting personnel and supplies. The outcomes of these purchases directly affect the Air Force's ability to succeed in a contingency environment.

Purchasing goods and services to support contingency operations can provide several types of benefits to the U.S. Air Force. As with most types of outsourcing, contract support frees up airmen to perform core military activities, and providers that specialize in the outsourced goods or services can often offer improved performance and cost outcomes, if managed effectively. In addition, buying in theater reduces requirements for scarce transportation resources, potentially shortening deployment timelines, and garners host-nation support for U.S. military presence. And having the capability to purchase as needed, rather than being forced to predict requirements in advance, helps commanders meet emerging demands and changing requirements associated with the realities of war.

Since September 11, 2001, the Air Force has been involved in two significant contingency operations in the U.S. Central Command (CENTCOM) area of responsibility (AOR): Operation Enduring Freedom (OEF) in Afghanistan and Operation Iraqi Freedom (OIF) in Iraq. To take advantage of the depth of contingency contracting experience built during recent operations, the Deputy Assistant Secretary of the Air Force for Contracting (SAF/AQC) asked RAND Project AIR FORCE to gather and analyze data on goods and services purchased to support Air Force missions in OIF to determine (1) the size and extent of contractor support for OIF[1] and (2) how plans for and the organization and execution of contingency contracting activities might be improved to better support the warfighter in future operations. Creating a comprehensive database was a major task of this study, as such data were heterogeneous both within and across purchasing organizations. The motivation for this study was that such insights from comprehensive data on recent multiyear contingency contracting experiences would help inform decisions about a number of important policy issues.

First, such data could be used to improve the Air Force's ability to plan for combat operations at contingency operating locations, particularly by linking purchases to supplemental information about the phases of operations (e.g., deployment, the building of a base, the sustainment of operations at a base, the closing of a base) and mission activities supported by those purchases.[2] For example, the Air Force could make more informed trade-offs between purchasing required assets as needed during operations in theater and purchasing them in advance and then using airlift or other transportation assets to move

---

[1] U.S. Government Accountability Office (2005) notes the difficulties of determining how funding for the global war on terrorism (GWOT), which includes OEF and OIF, has been spent.

[2] According to the Air Force Federal Acquisition Regulation Supplement (AFFARS, Appendix CC, para. CC-502-1–CC-502-4), contracting support for deployments takes place in four phases: initial deployment, buildup, sustainment, and termination/redeployment.

materials from the continental United States (CONUS) or regional storage locations to operating locations.

Second, purchasing data could be used to improve training for future contingency contracting officers (CCOs). Insights about how purchasing evolves with operational phases could be used to design more realistic training courses.[3] Further, information about typical goods and services purchased, types of contracts used, and supply bases at specific locations could be used to better prepare CCOs who are getting ready to deploy.

Third, information about contracting workloads at different types of bases and other purchasing organizations during different phases of operations could be used to better align CCO organizations and personnel assignments (encompassing both numbers and skill levels) with warfighter requirements.

Finally, descriptive data on individual transactions are important inputs in efforts to improve purchasing practices across the theater. Such improvements might include more effective price negotiation based on improved visibility of prices of similar goods or services, as well as identification of potential opportunities to improve the Air Force's leverage with key suppliers through contract consolidation across commodity groups or sites.

Defining Contingency Contracting for OIF

AFFARS provides us with the following definitions relevant to this research. A *contingency* is "an emergency, involving military forces, caused by natural disasters, terrorists, subversives, or required military operations," and *CCOs* are people with "delegated contracting authority to enter into, administer, and terminate contracts on behalf of the Government in support of contingency . . . operations" (AFFARS, Appendix CC, para. CC-102).

---

[3] Although we did not systematically study training issues, anecdotal evidence from our discussions with CCOs suggests that the contingency contracting curriculum then offered at the Defense Acquisition University (DAU) did not fully prepare CCOs for their deployment responsibilities.

In this monograph, we use a broad definition of contingency contracting for OIF. We include war preparations in early FY 2003, the war itself in mid–FY 2003, and postwar activities beginning in the latter part of FY 2003. U.S. Central Command Air Forces (CENTAF) was the primary major command involved in Air Force operations; however, it was not the only one. Many other commands and organizations made purchases in support of this effort. For example, purchases were made to support U.S. Air Forces in Europe (USAFE) bases, Air Force Special Operations Command (AFSOC) forces, and Air Mobility Command (AMC) operations. (See Appendix B.)

CENTAF purchases to meet operational requirements can be categorized by the geographic location of the purchases, referred to within the CENTAF financial management community as either *rear* or *forward*. Rear purchases were made by functional groups with specific areas of expertise (e.g., force protection), located at 9th Air Force Headquarters at Shaw AFB, South Carolina, to address requirements common across Air Force operating locations within the CENTCOM AOR. Some of these—such as requirements for storage, maintenance, and refurbishment of war reserve materiel (WRM)—may have been known in advance; others, such as requirements for new technologies for force protection for all theater bases, occurred in real time. Forward purchases were made by deployed CCOs to support the day-to-day needs of troops and operating locations in theater.[4]

Table 1.1 provides summary data on the magnitude of expenditures associated with rear and forward purchases for FY 2003, as provided by CENTAF comptroller and functional area resource advisors.[5] The details of the types of data included in the forward CCO purchases are discussed in the next chapter.

---

[4] Army contracting officer MAJ Ruthann Haider notes that CCOs "supplement the military supply system by providing deployed commanders a means to obtain the needed materials, services and supplies not readily available through normal supply channels" (Edwards, 2002).

[5] The functional areas included in Table 1.1 represent the primary areas that purchased goods and services for forward operating locations.

Table 1.1
CENTAF Rear and Forward Purchases for FY 2003

Purchaser	Expenditures Captured in Our Database ($ millions)
CCOs (forward)	OIF bases and other purchasing organizations
FY 2003	246
Functional areas (rear)	CENTCOM AOR, using GWOT funding
Civil engineering (FY 2003)	309
Logistics (FY 2003)	296
Communications (FY 2003)	206
Force protection (FY 2003)	55
Operations (FY 2003)	30
Total	1,142

NOTE: Purchases were made using operations and maintenance (O&M) funds.

To put these contingency purchases into a broader spending context, we examine total Air Force OIF expenditures for FY 2003–FY 2004 (see Table 1.2). These include a wide range of costs associated with personnel, personnel support, operating support, and transportation for both active duty and reserve components.

Table 1.2
Air Force Expenditures for OIF

Fiscal Year	Reported Obligations for OIF ($ billions)
2003	7.9
2004	6.6

SOURCE: Data provided to RAND by Defense Finance and Accounting Service (DFAS).

NOTE: Some types of expenditures were reported for the GWOT as a whole. In these cases, estimates were made for the portion applying to individual contingency operations such as OIF. We thank John Johnson and Capt Joshua Weed of the Air Force Financial Management and Budget Office (SAF/FMBO) for their assistance with these data.

Research Approach

To develop a baseline of contingency contracting for OIF and obtain insights relevant to the policy issues discussed earlier, we sought to develop a comprehensive database of Air Force purchases associated with OIF. In the analyses that are documented in this monograph, we focus on CENTAF purchases made by CCOs at locations within the theater of operations, i.e., forward purchases. (See Appendix B for a discussion of additional data relevant to purchases for OIF.) We chose these data for several reasons. The current lack of visibility into the details of the forward transactions and the decentralized nature of the CCO purchases suggest that there should be opportunities to improve planning for and execution of these activities, e.g., through preplanning for certain types of goods or services, more effective price negotiation, or contract consolidation with key suppliers to the AOR. In addition, the numbers of dollars and individual transactions for CENTAF are much greater than equivalent data received from other commands and organizations that supported OIF.[6]

Our database allows us to analyze the CENTAF CCO purchases in several important ways. First, we can examine purchases occurring at individual operating locations over time. With additional data on the evolution of activities at individual bases, we can characterize purchasing activities by phase of operations, i.e., initial deployment, buildup, sustainment, and termination and redeployment. Information about operational missions, such as the types of aircraft supported at individual locations, would provide additional context. Looking across bases in the region, we can analyze purchases according to characteristics of the bases themselves, such as size (e.g., classified by

---

[6] Data on CCO purchases to support USAFE bases would provide an interesting point of comparison to the CENTAF data discussed in this monograph. However, we were unable to find a source of aggregated data similar to the CENTAF databases used in this analysis. Examination of the disaggregated USAFE data was infeasible with resources available for this study.

base population or geographic dispersion) or bases known to be temporary operating locations versus those treated as more permanent.[7]

Second, we can characterize purchases of different types of goods and services. In addition to basic descriptions of dollars spent and numbers of purchases, we can identify which goods and services were purchased across multiple locations and those with limited use of local suppliers (suggesting limited local supply bases).

Third, our data can be used to characterize the set of suppliers associated with theater purchases. In particular, we can identify those suppliers with which the Air Force did a great deal of business, the types of goods and services purchased from them, the locations serviced by them, and so forth. Supplier names can be used to gauge (though imperfectly) whether suppliers are U.S. firms or located within the AOR. Such information can be valuable when negotiating host-nation agreements on the presence of U.S. military.

Fourth, our data can be used to identify selected types of purchasing tools used to procure categories of goods and services. In particular, we can identify the extent of use of government purchase cards (GPCs) and blanket purchase agreement (BPA) contracts.[8]

Finally, our database can be used as a resource for CCOs who are looking for examples of how to structure specific types of contracts or to examine approximate price ranges for purchases. For example, a CCO who has to write a new contract for water or food services at a deployed location could identify similar contracts at other locations and then contact the appropriate personnel there to obtain relevant samples.

---

[7] We sought supplemental data on types and levels of activities at individual locations for our analyses but were unable to obtain them.

[8] BPA contracts are used to satisfy anticipated recurring requirements for goods or services. They are designed to reduce transaction costs and speed up the procurement process "by establishing 'charge accounts' with qualified sources of supply" (AFAA, 2004). The contracts specify the range of goods and/or services covered by the agreement, price lists, total dollar limits, and time limits. Contracting officers (or other authorized and trained personnel) can then place "calls" against those agreements to meet specific user requirements that fall within the bounds of the agreements.

Preview of Findings and Recommendations

In this monograph, we demonstrate that it is possible, though currently difficult, to develop a database that allows detailed analysis of individual purchases for a contingency, including who purchased what goods and services, when, for what cost, from which firms, and how the transactions were accomplished. Such information is an important input in addressing policy questions related to combat support planning, the development of strategic relationships with suppliers, the effective use of different contract instruments, CCO assignments, and training for CCOs.

We recommend that the Air Force develop a standardized method for gathering detailed data on contingency purchases on an ongoing basis, as well as summarizing those data according to, for example, the types of goods or services purchased or the suppliers used, to facilitate such analyses in the future. Current efforts within CENTAF contracting are a positive step in this direction. Additional data, such as the characteristics of operating locations and types of mission activities, would provide the broader context needed to understand observed trends in the data.

The Air Force is in the process of reviewing current contracting organizations, including those overseas, to determine what future organizations should look like. This database could also help with that restructuring effort.

Structure of This Monograph

The remainder of this monograph is organized as follows. Chapter Two describes the methods used to build the database for the analyses presented in subsequent chapters. Chapter Three provides a baseline analysis of contingency contracting purchases for OIF. Chapter Four highlights findings and policy implications from the data analyses. Chapter Five presents our recommendations. Appendix A is a discussion of selected contingency contracting issues. Appendix B describes data sources in detail. Appendix C provides more details about how we prepared and processed the data.

CHAPTER TWO

Data and Methodology for Building the Database

In this chapter, we briefly describe the data we collected on CENTAF CCO purchases to support OIF and the methodology we used to clean and refine the data for our analyses. Appendixes B and C contain more details on data sources and processing methods.

Data

The analyses reported here are based on CENTAF data describing purchases made by deployed CCOs assigned to Air Force operating locations within the CENTCOM AOR. These include operational bases and other organizations, such as the Combined Air Operations Center (CAOC),[1] the Joint Contract Center (JCC) for Jordan, and Rapid Engineer Deployable Heavy Operational Repair Squadron Engineer (RED HORSE) teams.[2] Henceforth, we will refer to this group of bases and organizations as *purchasing organizations*. When we began our study in FY 2003, CCO data collected by the Air Force contracting community were decentralized and heterogeneous. Individual CCOs were expected to maintain purchasing logs describing each purchase they

---

[1] The CAOC provides a planning and command-and-control capability for the entire theater. At the height of recent operations, 1,966 individuals were working within CAOC facilities (Moseley, 2003, p. 3).

[2] RED HORSE teams are mobile, modular teams of engineers trained and equipped to meet a wide range of construction and repair requirements in a contingency environment, e.g., runway repair, force beddown construction (Worrell, 2004).

made during deployment (so-called *PIIN logs*, for the procurement instrument identification number used to identify transactions). Typical deployments spanned three or four months to a year.[3] Although some individual commands had developed guidelines for the logs, the Air Force did not provide a standard format.

Contracting officers have the sole authority to spend money on behalf of the Air Force to satisfy warfighter requirements; however, they do not work alone. Contracting officers work hand in hand with financial management personnel who are responsible for verifying that purchases are appropriate for certain types of funding, that funding is available, and that the purchasing organization has the authority to use those funds for the requested purchase.

Because of these responsibilities, the CENTAF comptroller maintains large databases of Air Force purchases occurring at all Air Force purchasing organizations within the CENTCOM AOR. These databases contain details about each organization's purchases by fiscal year. See Table 2.1 for a list of the primary data elements tracked for each purchase. The databases are organized by the type of contingency funding used to pay for the goods or services and by the type of payment, i.e., contract or GPC.

The funding category is denoted by a funding code, called an ESP (emergency and special program) code, corresponding to a specific budget appropriation for O&M funds associated with individual contingency operations.[4] For example, during the past several years, CENTAF has used funds designated for Operation Southern Watch (OSW), OEF, and OIF.[5]

---

[3] During our study time frame, the typical rotation schedule for contracting officers was lengthened from three to four months. A small number of "permanent party" personnel was assigned to selected locations for a full year.

[4] O&M funds are also referred to as *3400 money* because of the number associated with the appropriation category. DFAS publishes the *DFAS Manual 7097.01.* Chapter 3-AF-ESP.4.11 of that manual lists ESP codes for the Air Force. (See DFAS, 2003.)

[5] The ESP codes for OEF and OIF are 7C and ZA, respectively. These can be found in DFAS (2003). Two other codes relevant for our research are VA and YC. The VA code is listed as "rescinded" in the manual, but, according to a Robins AFB history account, VA

Table 2.1
Important Data Elements Tracked in CENTAF Comptroller Databases

Data Field	Description
Cardholder	For GPC only, the person authorized to make the purchase
Date of request	Date the purchase request was received from the user
Committed amount	Estimated cost of the purchase for GPC; any remaining commitment for contract transactions[a]
Obligated amount	Actual cost of the purchase
Company name	Name of the supplier
EEIC	Element of expense investment code[b]
Detailed item description	Description of goods or services purchased
Number of items	Quantity purchased
Unit of issue	Quantity of units for the purchase (e.g., liters, bottles, months, "each")
Not applicable, Form SF44, BPA, call number, or partial pay	Description of the payment mechanism used for the purchase
Requesting organization	Organization requesting the goods or services purchased
Contract number	Unique 13-digit contract identifier
Date paid (for contracts), purchase date (for GPC)	Date of payment
Comments	Additional information provided at the discretion of the contracting officer

[a] The expected cost for a contract is noted once the contract is put in place. As dollars are actually spent against the contract, the committed dollars are reduced, and eventually eliminated, once the contract has been fully executed.

[b] The Summary of Major Changes to U.S. Department of Defense (DoD) Financial Management Regulation 7000.14R, Vol. 9, Definitions (May 2005), states that EEICs are five-digit codes "designed for use in budget preparations and accounting systems to identify the nature of services and items acquired for immediate consumption (expense) or capitalization (investment)" (DoD, 2005, p. 4).

was the ESP code for OSW and YC was the code for Operation Northern Watch (ONW) (Robins AFB and Warner Robins Air Logistics Center, 2001, Chapter Five, Table V-5).

For each funding code used by a given purchasing organization, the CENTAF comptroller maintains separate databases for purchases made through some type of contract vehicle and purchases paid for by a GPC.[6] For example, a base authorized to use both OIF and OEF funding codes will have as many as four tracking files for each fiscal year: OEF purchases made via GPC, OEF purchases made via contract, OIF purchases made via GPC, and OIF purchases made via contract.

When we began our study, we planned to use the OIF funding code to isolate purchases associated with OIF. However, the CENTAF comptroller's office advised us that this distinction is problematic. The timing of both OEF and OIF, and the close proximity of operations, made it extremely difficult to disentangle exactly which purchases were made to support which operation. Many purchasing organizations within the AOR during FYs 2003 and 2004 were supporting operations in both Iraq (for OIF) and Afghanistan (for OEF) and, as a result, utilized both types of funds. (However, this is not a problem for selected bases established solely to support one of these operations, e.g., bases in Iraq that supported only OIF.) Instead of using the OIF ESP code, the CENTAF comptroller recommended that we define OIF purchases as those occurring at purchasing organizations primarily supporting OIF during this time. These were identified by the CENTAF comptroller as those within Bahrain, Egypt, Iraq, Jordan, Kuwait, Oman, Qatar, Saudi Arabia, and the United Arab Emirates. Organizations within the "-stans" (Kyrgyzstan, Tajikistan, Turkmenistan, Pakistan) were excluded, as they were primarily supporting OEF.

Completeness

The CENTAF comptroller provided databases on each of the purchasing organizations supporting OIF in both FY 2003 and FY 2004. Table 2.2 lists the 24 organizations (and their locations) for which we have data. In most cases, these databases represent all available data on

---

[6]   Contract data files are called *616 files*, named for the authorization form (Form 616) used to establish purchasing accounts (likened to checkbooks).

Table 2.2
Purchasing Organizations Included in Our Database

Purchasing Organization	Country
Al Dhafra AB	UAE
Al Jaber AB	Kuwait
Al Minhad AB	UAE
Al Udeid AB	Qatar
Ali Al Salem AB	Kuwait
Ar'Ar Airport	Saudi Arabia
Azraq AB	Jordan
Baghdad International Airport	Iraq
Balad AB	Iraq
Cairo West AB	Egypt
Camp Snoopy	Qatar
King Abdullah AB	Jordan
King Faisal AB	Jordan
Kirkuk AB	Iraq
Masirah AB	Oman
Prince Sultan AB (PSAB)	Saudi Arabia
Seeb International Airport	Oman
Shaikh Isa AB	Bahrain
Tabuk AB	Saudi Arabia
Tallil AB	Iraq
Thumrait AB	Oman
CAOC	Saudi Arabia, Qatar
JCC	Jordan
RED HORSE	Multiple locations

CCO purchases at those locations. However, seven of these recorded some or all of their contract transactions during this period in a centralized electronic database called the BQ system, rather than in the financial management spreadsheets.[7] These purchasing organizations include Al Dhafra AB, United Arab Emirates; Al Jaber AB, Kuwait; Al Udeid AB, Qatar; Ali Al Salem AB, Kuwait; CAOC, also at Al Udeid AB, Qatar; PSAB, Saudi Arabia; and Seeb International Airport, Oman. Although we were given information about the dollar amount of purchases recorded in BQ, the BQ data do not provide detailed descriptions of these purchases. In addition, we do not know the number of transactions associated with the dollars in the BQ system. Because data for these locations are incomplete, encompassing only GPC expenditures in some cases, we are unable to include them in some of the analyses in this document.

In addition, we learned during a visit to Al Udeid AB that some urgent base-support purchases early in the war effort were made through the Air Force's WRM contract rather than through the regular CCO purchasing process. Thus, these data were not captured in the CENTAF comptroller databases used in our analyses. While we do not believe that the use of the WRM contract for these types of purchases was extensive, we were unable to assess its magnitude.[8] These issues illustrate the complexities of creating a truly comprehensive, detailed database of CCO purchases given current contracting systems and practices.

Time Frame

Figure 2.1 provides information on the time frames for purchasing activity for each of the OIF purchasing organizations during FYs 2003 and 2004. (Purchasing activity corresponds to operations for each of these organizations.) Only five had contracting activity throughout both years. Some were active for only a few months.

---

[7]   The BQ system is the U.S. Air Force's Standard Base Level General Accounting and Finance System. Its structure and use are described in DFAS (2000).

[8]   We thank the WRM site manager at Al Udeid for bringing this to our attention.

Figure 2.1
Timelines for Purchasing Activity, by Purchasing Organization

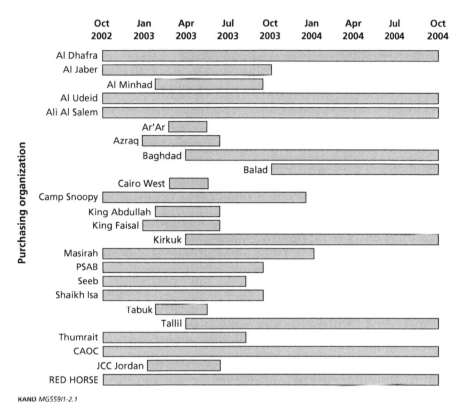

RAND MG559/1-2.1

Database Construction

The CENTAF comptroller data on CCO purchases were tracked in Microsoft® Excel® spreadsheets. Data fields and spreadsheet formats were similar, but not identical, for contract and GPC files across FYs 2003 and 2004.

We developed a detailed process to merge these files into an aggregated master database that we could use for our analyses. As part of the process, we reviewed and corrected several variables, including dates associated with each purchase and information related to contractors. See Appendix C for the details of this process.

We also created three new variables for our analyses. First, we created a variable for the purchasing organization (i.e., the base or other organization) with which the comptroller associated the transaction. Second, we used the text description for each transaction to categorize the purchase according to one or more types of good or service. Table C.1 in Appendix C provides a list of the categories we created, as well as a general description of the types of purchases included in each. And third, we created a variable for the type of transaction. This variable identifies whether the purchase was made using a GPC or a contract vehicle. Contracts are further broken down into BPAs and "other" contracts.

Table 2.3 displays the data fields that we used in the analyses described in subsequent chapters of this monograph.

Table 2.3
Data Fields Important for Our Analyses

Category	Description
Purchasing organization	Base or other organization
Date of request	Date purchase was requested
Purchase category	RAND-created variable based on a description of the purchase
Obligated dollars	Actual price paid
Supplier	Firm providing the purchased goods or services
Type of transaction	Purchasing tool used; categories for our analyses are GPC, BPA, and other type of contract

CHAPTER THREE

Baseline of Contingency Contracting for OIF

In this chapter, we provide the results of our baseline analysis of purchases supporting Air Force activities during OIF. Specifically, we describe CCO purchases occurring during FYs 2003 and 2004 at the Air Force operating locations listed in Chapter Two.

The analyses are organized as follows. After an overview of expenditures, we describe

- *who* (which organizations) made purchases
- *what* types of goods and services were purchased
- *when* the purchases were made (time periods)
- *how* the purchases were made (contracting tools used)
- *from whom* (suppliers) the purchases were made.

Overview of Expenditures

By fiscal year, the total obligations (in dollars) and the number of transactions in our CENTAF CCO database are shown in Table 3.1. As noted in Chapter Two, we are missing contract data for seven purchasing organizations at varying periods of time during FYs 2003 and 2004. The totals in Table 3.1 include the partial data we have on these seven organizations.

Table 3.1
Transactions and Obligations for FYs 2003 and 2004
Captured in Our Database

Fiscal Year	Transactions (n)	Obligations ($)
2003	20,480	245,558,938
2004	17,819	125,970,515
Total	38,299	371,529,453

In many of the analyses presented in this chapter, we exclude the seven organizations for which we have only partial information—Al Dhafra, Al Jaber, Al Udeid, Ali Al Salem, CAOC, PSAB, and Seeb—focusing on those for which we have all available data on CCO purchases. The transactions and obligated dollars associated with the 17 "complete" organizations are summarized in Table 3.2.

A comparison of Tables 3.1 and 3.2 shows that we are excluding approximately 14,000 transactions totaling more than $60 million. These represent more than 36 percent of the transactions but less than 17 percent of dollars spent. The excluded data primarily represent GPC transactions.

Table 3.2
Transactions and Obligations for FYs 2003 and 2004 for Purchasing
Organizations with Complete Information

Fiscal Year	Transactions (n)	Obligations ($)
2003	13,570	212,677,724
2004	10,692	97,181,578
Total	24,262	309,859,303

NOTE: Dollar amounts do not sum due to rounding.

Who

Figure 3.1 shows how the level of expenditures varied across the 24 purchasing organizations captured in our database during FYs 2003 and 2004. Although we are missing detailed data on transactions recorded in the BQ system, the CENTAF comptroller provided us with information about the level of these expenditures. The horizontal bars in Figure 3.1 break the spending at each organization into four pieces: CCO spending in FY 2003, missing BQ spending in FY 2003, CCO spending in FY 2004, and missing BQ spending in FY 2004. Organizations are sorted by total CCO purchases during FYs 2003 and 2004.

Figure 3.1
Spending at OIF Purchasing Organizations, FYs 2003 and 2004

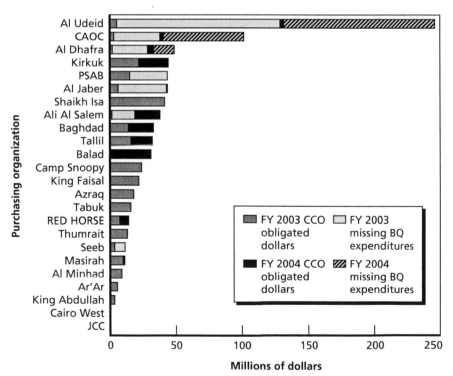

The most spending by far occurred at Al Udeid, which includes both the air base operations (labeled "Al Udeid") and the CAOC, which relocated from PSAB to Al Udeid during this period. Al Udeid is, and is expected to remain, the Air Force's primary operating location in Southwest Asia. The location also served as the forward headquarters of the Air Force in Southwest Asia during both OIF and OEF. This may offer an explanation for the extensive spending at Al Udeid. Unfortunately, Al Udeid's and the CAOC's contract expenditures were captured only in the BQ system during this period, and, as previously discussed, the lack of resolution in the BQ database prevents detailed analysis.

In addition, we would like to link RED HORSE purchases supporting construction or repair projects to individual locations benefiting from those purchases. However, the CENTAF databases do not provide that level of visibility into the RED HORSE transactions.

As seen in Figure 2.1 in Chapter Two, the periods of contracting activity captured by our data vary across purchasing organizations. To control for these differences, we examined the average monthly spending for organizations in Figure 3.2, based on both CCO spending captured in our database and missing BQ spending. The average monthly expenditure was $2.6 million during FYs 2003 and 2004. Fifteen of the 24 organizations spent between $1 million and $4 million per month during this period. Al Udeid stands out as the highest-spending organization at more than $10 million per month.

In the analyses below, unless otherwise noted, we include only the 17 organizations for which we have complete data; we exclude the seven organizations for which we have incomplete data due to the limitations of the BQ system—Al Dhafra, Al Jaber, Al Udeid, Ali Al Salem, CAOC, PSAB, and Seeb.

What

Deployed CCOs purchased a variety of products to support OIF operations during FYs 2003 and 2004. As described in Appendix C, to analyze these purchases, we created 45 categories of goods and services

Figure 3.2
Average Monthly Spending by OIF Purchasing Organizations,
FYs 2003 and 2004

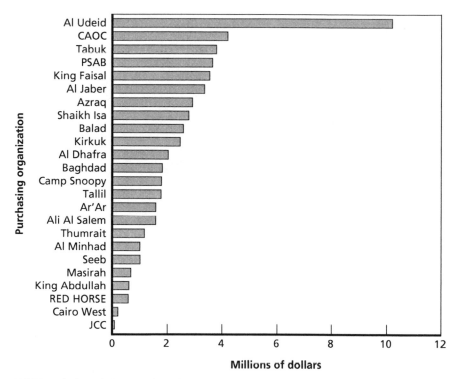

NOTE: Includes missing BQ data for the seven affected organizations.
RAND MG559/1-3.2

and used a computer program to assign transactions to these categories based on key words found in the text descriptions of the purchases.

Figure 3.3 shows how spending in FYs 2003 and 2004 was distributed among these categories, with the categories ordered from most to least spent on the particular good or service. In many cases, the description of a purchase fit clearly into only one category. Other transactions included purchases of more than one disparate item or items that were ambiguously described and might, because of the use of key words in the program, fit into more than one category. For example, the text description might include a laptop computer (computer equipment) and a printer (office supplies and equipment), or the purchase may be described as a "desk for chapel," which could be interpreted by the computer program as furniture (the desk) or morale, welfare, and recreation (MWR) (the chapel). The darker portions of the horizontal bars in Figure 3.3 show obligations for transactions that included items in only one category ("single"); the lighter portions show obligations for transactions that were assigned to more than one category ("multiple"). The figure shows that descriptions of transactions in categories such as food service (toward the top of the figure), water (a third of the way down), and laundry services (toward the middle) were usually well specified and included items that belong only in those categories. On the other hand, most transactions involving purchases in the tools category (a quarter of the way down) also included items that our program assigned to other categories. We were unable to categorize less than $10 million of the more than $300 million in purchases during FYs 2003 and 2004; these purchases were placed in the "unknown" category (bottom of the figure).

The top three spending categories—construction supplies, vehicles, and construction services—were the same whether measured by single or multiple category obligations. Overall, the most dollars were assigned to the construction supplies category, but, for transactions assigned to only one category, more dollars were assigned to vehicle leases and construction services.

The information on spending shown in Figure 3.3 can assist in the analysis of new approaches to strategic spending. For example, many types of vehicles are pre-positioned as part of WRM. These vehicles are generally meant for short-term use, to be returned to WRM stock in preparation for the possibility of their need elsewhere. Replacing these vehicles may explain some of the spending on vehicles shown in Figure 3.3. A strategic planner, noting the size of these expenses over the long term, may find it useful to compare the costs of renewing short-term leases to the cost of prepurchasing greater numbers of vehicles, storing them, then shipping them to where they are needed in

Figure 3.3
Obligations by Category, FYs 2003 and 2004

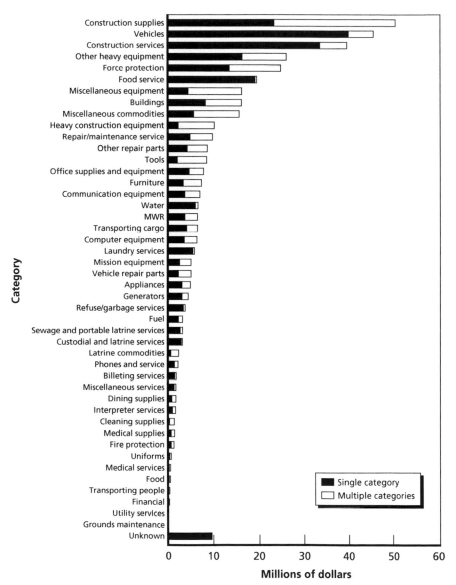

the theater.[1] Additional data would be required to evaluate the alternative approach.

Figure 3.4 displays the number of transactions recorded for each category of purchase, again showing "single" and "multiple" designations for the transactions. Construction supplies, miscellaneous commodities, and office supplies and equipment represent the largest number of transactions.

We can divide obligated dollars by the number of transactions (taking into account only single-category transactions) to compute the average transaction amount for each of the categories. Table 3.3 illustrates the variation in transaction amounts across categories. The average transaction for construction services was almost $200,000 but only $545 for transportation of personnel.

Finally, Figure 3.5 illustrates how the percentage of expenditures devoted to top categories of purchases varied across the 17 purchasing organizations for which we have complete data. The vertical lines represent the maximum and minimum percentages for each category, and the dot represents the average percentage. Categories are sorted in order of decreasing average percentages. To calculate the percentages for each organization, we divided the total obligated dollars associated with each category by the sum of total obligated dollars across categories for the organization (this includes double counting for transactions that are included in more than one category). A large ($21 million) transaction in FY 2003 at Shaikh Isa, representing 56 percent of total expenditures at that location, is reflected in the upper bound of the construction service range.[2]

---

[1]   Interestingly, an analysis of RED HORSE expenditures shows that $3.4 million of its $13.5 million in total purchases was spent on equipment. These purchases and leases included heavy construction equipment similar to that which RED HORSE units own. These units appear to have chosen to obtain needed equipment in theater rather than utilize scarce transportation resources to move equipment from storage locations.

[2]   We also calculated the median percentage for each of the categories, since the averages of such a small sample are affected by outliers such as the Shaikh Isa construction project. The medians are slightly lower than the averages—10 percent for construction supplies, 8 percent for vehicles, 2 percent for construction services, 4 percent for other heavy equipment, 0 percent for food service, and 5 percent for force protection—indicating that, for most organizations, these categories' percentages of expenditures were less than the computed averages.

Figure 3.4
Total Transactions by Category, FYs 2003 and 2004

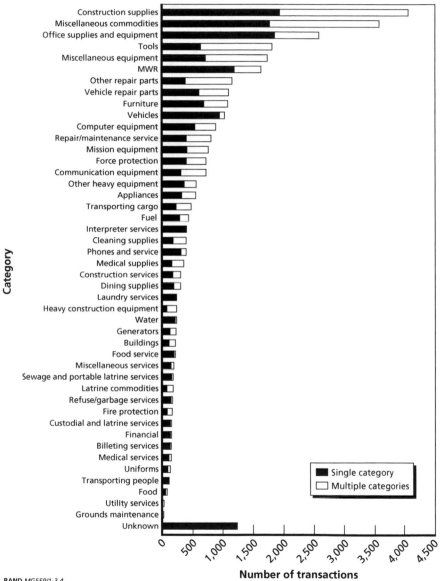

RAND *MG559/1-3.4*

Table 3.3
Average Transaction Amount for Selected Categories of Purchases

Category	Average Transaction Amount ($)
Construction services	196,323
Vehicles	41,538
Heavy construction equipment	29,790
Billeting services	10,468
Grounds maintenance	2,595
Transporting people	545

NOTE: Excludes transactions assigned to multiple categories.

Figure 3.5
Expenditures by OIF Purchasing Organizations Associated with Selected Categories of Purchases, FYs 2003 and 2004

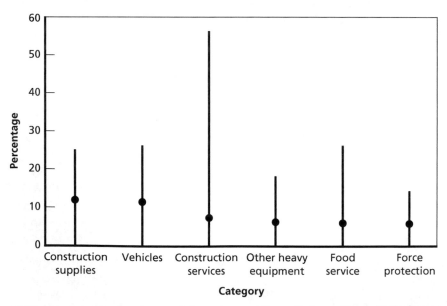

NOTE: The percentages are essentially unchanged if we eliminate JCC and RED HORSE to focus only on the bases.

RAND *MG559/1-3.5*

Supplemental information about the locations and activities supported by these purchases would be needed to explain the observed differences in expenditures across organizations.

When

Our database also allows analysis of purchases over time. Figure 3.6 shows that CCO purchases and transactions at these purchasing organizations were higher in FY 2003 than in FY 2004. This could be associated with the decline in the number of active bases (see Figure 2.1 in Chapter Two) or any number of other factors discussed in this section.[3]

Figure 3.6
Obligations and Transactions by Month, FYs 2003 and 2004

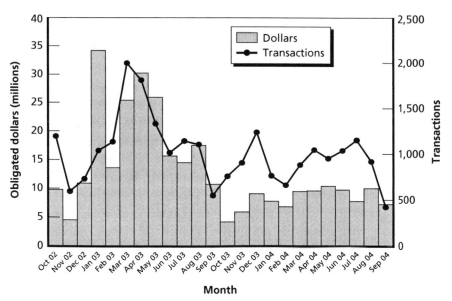

RAND MG559/1-3.6

---

[3] One of our interviewees suggested that the Air Force might have been able to reduce the level of base support it provided to personnel from other service branches during FY 2004.

We can disaggregate these data to examine how the level of expenditures varied over time at individual bases. Such data can be used to make comparisons across locations according to characteristics such as base population, types of operational missions (e.g., special operations, F-16s), existing base infrastructure, or permanency of the operating location.

For example, one might expect that initial expenditure patterns would vary a great deal depending on whether a base was a temporary operating location—i.e., one from which the Air Force planned to operate only for a limited time—or more permanent—i.e., one from which the Air Force planned to operate for an extended period. Intuition suggests that the Air Force may make more costly infrastructure investments at permanent locations that would not be cost-effective at temporary ones. In Figures 3.7 and 3.8, we examine expenditure

Figure 3.7
Total Obligated Dollars for Eight "Temporary" Locations

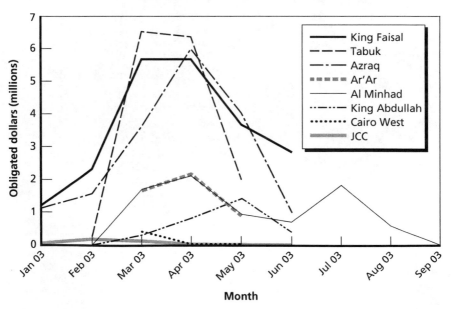

RAND MG559/1-3.7

Figure 3.8
Total Obligated Dollars for Four "Permanent" Locations

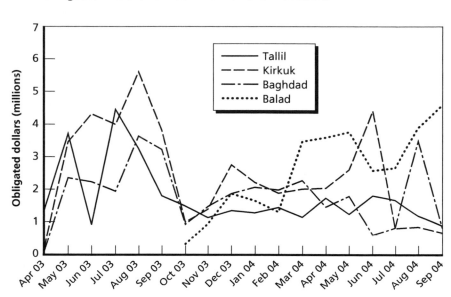

trends for eight "temporary" locations and four "permanent" locations, respectively, as identified by the CENTAF comptroller.[4]

Figure 3.7 shows the initiation, rise, and termination of obligations over time for eight "temporary" locations—Al Minhad, King Abdullah, Ar'Ar, Azraq, Cairo West, JCC, King Faisal, and Tabuk. There is a good bit of heterogeneity within this group, as might be expected, given the uncertain nature of operations at these locations. The sharp rise and fall of spending at King Faisal, Tabuk, and Azraq differ markedly from the patterns of other bases. These locations have not been utilized extensively by U.S. forces in the past and thus may have required more initial improvements to achieve the level of capability needed to accommodate forces.

---

[4]   RED HORSE expenditures are excluded, as they are not tied to specific operating locations. CENTAF personnel did not indicate whether the other four locations were temporary or permanent.

In Figure 3.8, we see expenditure trends for four "permanent" bases. Three of these locations, Baghdad, Kirkuk, and Tallil, show peaks in purchases during the first five months of operations, followed by a reduction and leveling out of expenditures. This pattern is consistent with a high expenditure level early in the life of a base, followed by a leveling out as the base transitions to sustained operations. Balad purchases (which did not begin until October 2003) look different, with purchases generally increasing over the entire year after the base's opening. Interestingly, the spending peaks of three temporary locations, Azraq, King Faisal, and Tabuk, equal or exceed the expenditure-month peaks of the permanent bases.

We can also examine expenditures at these four permanent locations by groups of months to approximate the evolution of activities at a base as it sets up, ramps up to operations, and sustains those operations. In Figure 3.9, we group expenditures in three-month intervals and examine expenditures across the bases, noting the maximum (top of the line), minimum (bottom of the line), and average expenditure (dot) for a base in each three-month interval.[5] Expenditure months are aligned by spending activity rather than the actual passage of time. That is, the first month with expenditures is counted as month one for a location, regardless of the actual calendar month and year.[6] The CENTAF comptroller suggested to us that such analyses can help inform programming decisions for future bases. For example, the experiences of these bases suggests that it would cost approximately $6 million to set up and run a new "typical" permanent base for the first three months. One would then expect the second three months of operations to be more expensive, requiring close to $10 million. After the first six months, the Air Force should expect such a location to require approximately $6 million to $7 million each quarter to sustain operations. Characteristics of the base and its operations may shed

---

[5]   Because of the small sample size, we calculated medians for each quarter as well. These were very close to the means.

[6]   The first month for Balad was October 2003. The other three bases began spending in April 2003, as can be seen in Figure 3.8.

Figure 3.9
Minimum, Maximum, and Average Quarterly CCO Expenditures at
Four "Permanent" Locations During FYs 2003 and 2004

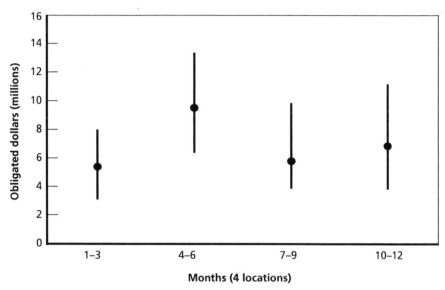

NOTE: Months are aligned by spending activity rather than calendar months. If we include expenditures for temporary bases in our analyses for months 1–3 and months 4–6, the average for each is virtually the same, but the maximum-minimum spread is much greater.

RAND MG559/1-3.9

light on observed differences in expenditure levels over time and thus assist with future programming decisions.

Time trends in base-level expenditures raise the question of how the goods and services purchased by CCOs might be changing over time. Interestingly, we find that purchasing requirements differ among operating locations over the same period, raising questions about the drivers of these differences. Figures 3.10 and 3.11 display obligated dollars for nine categories of purchases at Baghdad and Tallil, respectively, from April 2003 to September 2004. These two bases were chosen because of their similarities: Both were new bases established in Iraq in April 2003 with the same time frame for purchasing activity and

Figure 3.10
Purchases by Category at Baghdad

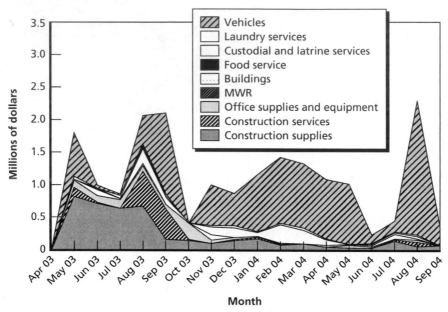

RAND *MG559/1-3.10*

comparable overall levels of CCO expenditures. The vertical axis for spending is on the same scale in the two figures. Tallil had much higher spending for construction supplies and construction services than did Baghdad at the initiation of the bases and over time as they became more established. On the other hand, Baghdad had much larger initial and continuing obligations for vehicle leases.

While our database alone cannot address underlying causes for the observed differences in spending patterns across locations over time, an analyst with additional information about conditions at these bases could use these data to help inform the resource decisions Air Force leaders will face in future operations. With location-specific information, such as base population, numbers and types of aircraft, types of missions, types and maturity of base infrastructure, geographic dispersion of facilities, and service branch responsible for base operating support (BOS), analysts could perform more sophisticated

Figure 3.11
Purchases by Category at Tallil

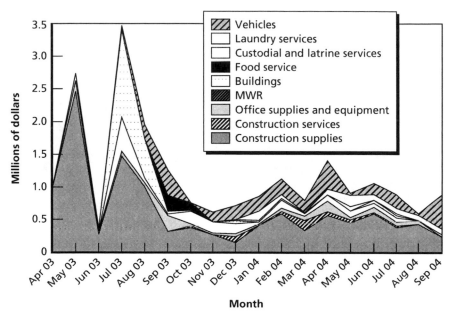

evaluations to determine the correlation between these factors and spending patterns over time.[7] The results of such analyses could be used to make programming decisions about new bases, plan for transportation requirements, match CCO resources with user requirements, and so forth. We return to this in Chapter Four.

How

CCOs have a variety of instruments with which to make purchase payments. Our data allow us to identify two particular types of instruments for further analysis: GPCs (essentially government-issued credit cards) and BPAs. We first compare purchases made using GPCs to

---

[7] Such information would need to be dynamic due to the fluid nature of wartime operations.

purchases made through contract instruments that are recorded in CENTAF comptroller 616 files. As shown in Figure 3.12, GPC purchases represented over a third of the transactions made in FYs 2003 and 2004, but they represented less than one-tenth of the dollars spent. Since GPCs are designed for purchases of small items, such as office supplies—many of which can be made over the Internet—this is an understandable finding.

Generally, contract transactions were much larger than GPC transactions, as shown in Table 3.4.

Figure 3.12
GPC Versus Contract Purchases in FYs 2003 and 2004

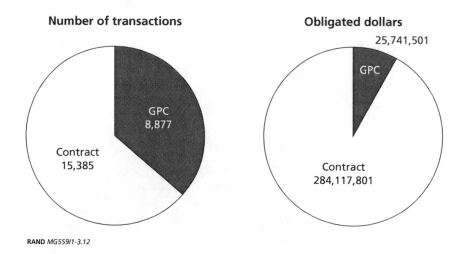

RAND MG559/1-3.12

Table 3.4
Average Size of GPC and Contract Purchases

Type of Purchase	Obligations ($)	Transactions (n)	Average Transaction Amount ($)
Contract	284,117,801	15,385	18,467[a]
GPC	25,741,501	8,877	2,900

[a] This includes the $21 million transaction at Shaikh Isa for a ramp extension. If it is excluded, the average contract transaction is approximately $17,000.

The relative use of GPCs and contracts for purchases varied by purchasing organization, as is clear in Figure 3.13. Here we include the organizations with obligations recorded in the BQ system, as well as those for which we have complete data. Note that the CCO data collected by the CENTAF comptroller for Al Udeid and the CAOC are GPC only; all contract data were in the BQ system for those two locations.

Figure 3.13
GPC Versus Contract Obligations, by Organization, FYs 2003 and 2004

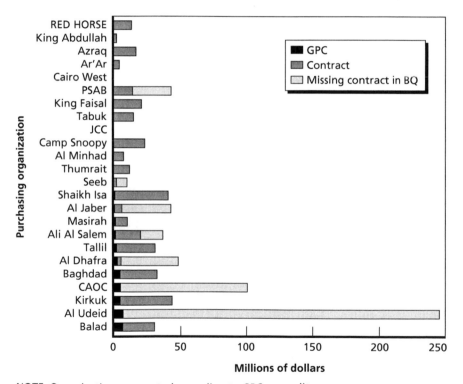

NOTE: Organizations are sorted according to GPC expenditures.

Figure 3.14 shows the use of GPC versus contracts for purchases in categories for which there was at least $100,000 in GPC purchases. These data include only the 17 organizations for which we have complete purchasing data.

The figure shows that the majority of GPC purchases were for goods (e.g., construction supplies, tools, office supplies); only a few services (such as repair or maintenance service and phone service) had significant GPC purchases. For four categories of purchases

Figure 3.14
GPC Versus Contract Purchases, by Category of Purchase

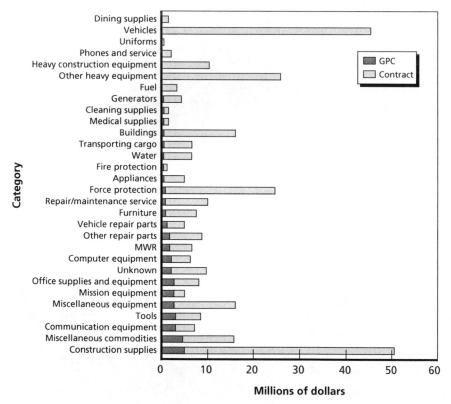

NOTE: Categories are listed in order of increasing GPC purchases. Category totals include purchases grouped into more than one category.
RAND MG559/1-3.14

(communication equipment, mission equipment, fire protection, and uniforms), GPC purchases represented more than 40 percent of total purchases.

Table 3.5 provides more detail about the differences between contract and GPC purchases. It lists the top 10 categories (and their percentage of expenditures) for both GPC and contract purchases. These top 10 categories represent approximately 75 percent of GPC purchases and 70 percent of contract purchases.[8] Interestingly, construction supplies rank first for both GPC and contract transactions. Other contract transactions were concentrated in more expensive goods and services, while GPC purchases were used primarily for smaller equipment and supplies.

Table 3.5
Top 10 Categories for GPC and Contract Purchases

Rank	GPC	Contract
1	Construction supplies (13%)	Construction supplies (13%)
2	Miscellaneous commodities (12%)	Vehicles (13%)
3	Communication equipment (8%)	Construction services (11%)
4	Tools (8%)	Other heavy equipment (7%)
5	Miscellaneous equipment (7%)	Force protection (7%)
6	Mission equipment (7%)	Food service (5%)
7	Office supplies and equipment (7%)	Buildings (4%)
8	Unknown (5%)	Miscellaneous equipment (4%)
9	Computer equipment (5%)	Miscellaneous commodities (3%)
10	MWR (5%)	Heavy construction equipment (3%)

NOTE: The percentages represent the percentage of GPC or contract spending associated with that category and include transactions assigned to multiple categories.

---

[8] The top 10 categories represent $28.7 million of the $38 million in GPC purchases and $248.8 million of the $360 million in contract purchases. Note that these totals include transactions assigned to multiple categories, so they differ from those in Table 3.4.

A 2004 Air Force Audit Agency report (AFAA, 2004) describes problems with the implementation of BPAs at Al Udeid. The problems included improper use of BPAs by personnel who lacked appropriately documented training, as well as inadequate oversight. Poor oversight resulted in the failure to deobligate unused funds, inaccurate payments to vendors, and improper purchases of goods or services that were not included in the price lists. Presumably, these problems were not unique to Al Udeid. (See Calbreath, 2005.) For this reason, we sought to examine the extent and nature of the use of BPAs by the organizations for which we have complete contract data, so we developed a methodology (described in Appendix C) to identify them within our database.

We first note that, as seen in Figure 3.15, BPAs were used for 30 percent of the obligations during FYs 2003 and 2004 at the 17 organizations for which we have complete CCO data.

BPAs were also used throughout the organizations of interest, as shown in Figure 3.16, which shows *contract* spending by BPA and non-BPA contract instruments. Azraq, Baghdad, and Masirah used BPAs for 50 percent or more of their contract purchases.

Figure 3.15
Breakdown of Dollars by BPA, GPC,
and Other Contract, FYs 2003 and 2004

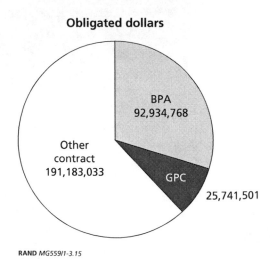

RAND MG559/1-3.15

Figure 3.16
Use of BPAs, by Organization

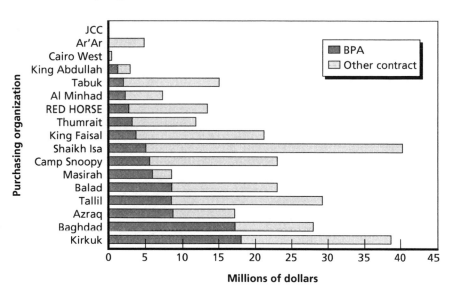

NOTE: Contract obligations only. Organizations are sorted by use of BPAs.
RAND MG559/1-3.16

In addition to their use throughout the region, BPAs were used to purchase goods or services in 40 of the 45 categories used in our analysis. Figure 3.17 shows BPA spending as a portion of total contract obligations for the categories that had more than $10 in BPA spending during FYs 2003 and 2004.[9]

Vehicles, other heavy equipment, and food service were categories for which BPAs were used frequently, with more than 60 percent of obligations associated with BPAs in each case. Water (79 percent), laundry services (56 percent), and refuse or garbage services (56 percent) also had high relative use of BPAs, though the dollar amounts spent were significantly lower.

---

[9]   Goods and services that had less than $10 in BPA obligations were financial services, food service, interpreter services, transporting people, utility services, and grounds maintenance.

Figure 3.17
Use of BPAs, by Category of Purchase

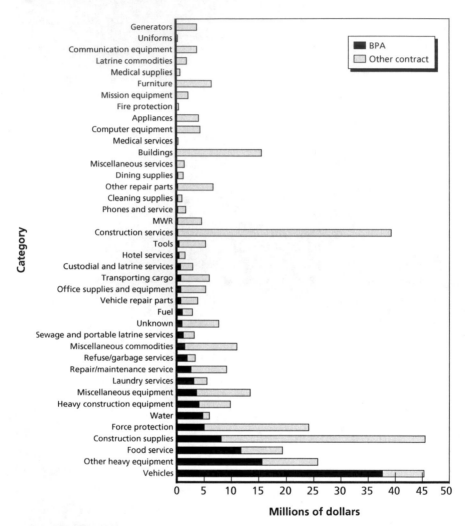

NOTE: Contract obligations only. Categories are sorted by BPA expenditures. Includes transactions counted in more than one category.
RAND MG559/1-3.17

Finally, while the significance of this is unclear, Figure 3.18 shows that the use of BPAs increased from FY 2003 to FY 2004: BPAs represented 25 percent of total obligations in FY 2003 and 41 percent of total obligations in FY 2004.

Figure 3.18
Proportion of BPA Spending in FYs 2003 and 2004

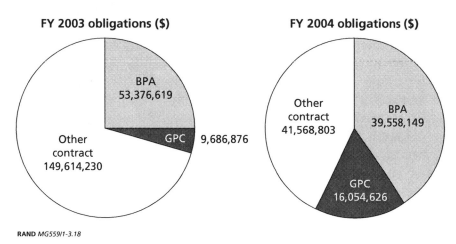

From Whom

Having examined who made purchases and what, when, and how the purchases were made, we now turn to the question of from whom goods and services were purchased. We examined the top 10 suppliers (in terms of dollars obligated) during FYs 2003 and 2004 by all obligations, for contract obligations alone, and for GPC obligations alone.[10]

The lists of top contractors overall and for contracts only include several so-called 10-percenters—companies that mediate the purchase of a good or service and then add a certain percentage (often 10 percent) as a fee. As a measure of the large presence of these companies, the top 10 suppliers are associated with more than 40 percent of the obligations during this period.

Based on firm names, the top firms by contract expenditures appear to be regional firms primarily, whereas GPCs were often used to make purchases from U.S. firms, presumably over the Internet. To get a better sense of what percentage of Air Force CCO purchases were

---

[10] We cannot list firm names here due to operational security considerations.

with regional firms, we examined the top 100 firms used in FYs 2003 and 2004, which represented 78 percent of the obligations during this period. Of these, 55 were regional firms.[11] Similarly, 59 of the top 100 firms for contract transactions were regional. However, the number was much smaller for GPC purchases, only 11 out of 100.

Interestingly, the rank of the suppliers in the CCO data changed from FY 2003 to FY 2004. Only two of the top 10 suppliers in FY 2003 were also in the top 10 in FY 2004.[12] While 10-percenters are important suppliers, this suggests that they do not dominate the supply base. Rather, supplier rankings appear to be sensitive to changes in theater purchasing over time—perhaps driven by changes in operating locations, changes in underlying purchases as military operations evolve, or other factors. Again, our data cannot answer these questions but can inform additional queries.

Top suppliers varied by purchase category as well. We examined the top five suppliers for three types of construction-related purchases: construction supplies, construction services, and heavy construction equipment. Interestingly, no firm appears in more than one of these three related categories. This could be due to segregation of the market, high levels of competition, or other factors.[13]

The top-ranked suppliers provided goods and services from a variety of categories. For each of the top five suppliers in FYs 2003 and 2004 (noted as firms A through E), Figure 3.19 displays the top five categories of purchases made through the supplier (with all other purchases counted in the bar labeled "Other"). Such diversity in lines of business is not surprising for 10-percenters.

---

[11] We determined this primarily through a Web search for firm ownership. Lacking other information, determination was based on the firm's name.

[12] In FY 2003, the top 10 suppliers accounted for $92.4 million, or 43 percent of total obligations. In FY 2004, the top 10 suppliers accounted for $45 million, or 46 percent of total obligations.

[13] Another hypothesis is that this pattern could reflect a desire to work with and provide business to a larger set of firms. More detailed knowledge of purchasing practices and the industrial base at specific locations would be needed to explain this pattern.

Figure 3.19
Top Five Purchase Categories for the Top Five Suppliers

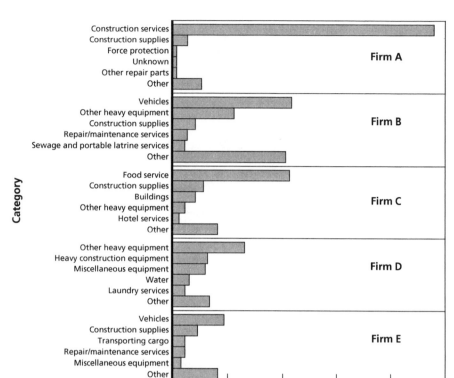

NOTE: One of the top categories of purchases from firm A consisted of items that our computer program found difficult to categorize and so placed in the "unknown" category.

RAND MG559/1-3.19

Top suppliers worked across multiple locations as well, as shown in Figure 3.20. In particular, firm E supplied goods and services not only in Iraq, but also in Qatar and Oman.

Such detailed information on suppliers' activities across the theater can assist CCOs in planning future acquisitions. For example, firm B and firm E both have contracts with Baghdad, Camp Snoopy, RED HORSE, and Tallil, and both companies provide goods and services within the same categories. While no contracts in our database

Figure 3.20
Purchasing Organizations Served by the Top Five Suppliers

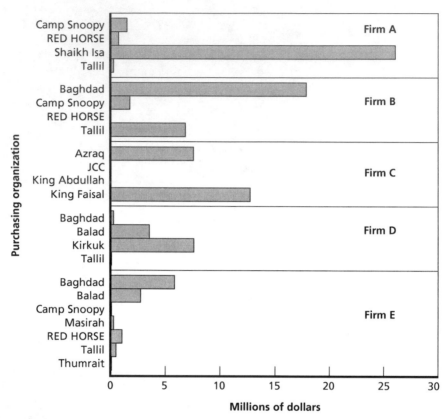

encompassed more than one purchasing organization,[14] there may be opportunities to increase leverage with providers by combining contracts across organizations and encouraging competition among providers. The data analyses presented here point to more detailed analyses that could inform such strategic purchasing decisions.

---

[14] We used the unique contract identification number to check for cross-organization contracts.

CHAPTER FOUR

Implications for Policy Issues

In this chapter, we use insights from the data and comments from interviews conducted during this project to address issues related to CCO staffing, CCO training, combat support planning, and the sharing of lessons within the theater.

CCO Staffing

Lacking hard data for detailed workload analyses, the Air Force traditionally has used general rules based on perceptions of past experience to determine how many contracting officers to allocate to deployed locations. This approach can lead to the need for adjustments after the fact to reflect real demands on CCOs' time.

One potentially important use of our database is to systematically assess workload for CCOs—measured in dollars obligated or transactions executed—across purchasing organizations. While neither measure is perfect (e.g., some small-dollar transactions may require more time and attention than do some big-dollar transactions), both measures are potentially important indicators of CCO time requirements.

We received supplemental data on CCO staffing for selected purchasing organizations for FY 2004 from CENTAF contracting. (See Table 4.1.) These data were obtained from the Automated Contract Tracking Tool (ACTT).[1] The staffing data include personnel on

---

[1] See Thrailkill (2003) and the discussion in the following chapter.

Table 4.1
CCO Staffing Data for FY 2004

Purchasing Organization	Average Number of CCOs in FY 2004
Al Dhafra	11
Al Udeid	14
Ali Al Salem	6
Baghdad	4
Balad	6
CAOC	4
Kirkuk	6
Tallil	6

90- or 120-day rotations, as well as the "permanent party" personnel (i.e., those deployed for a full year) at the CAOC, Al Udeid, Al Dhafra, and Ali Al Salem. Although personnel were rotating in and out every few months, these data represent the average number of CCOs in each of the organizations at any point in time during FY 2004.

For those organizations for which we have complete CCO data for FY 2004, we can compare the workload of contracting officers in terms of the average number of transactions per CCO and the average number of dollars obligated per CCO. This information is displayed in Figure 4.1.

Figure 4.1 indicates that there were large differences in CCO activities—as measured by obligated dollars and numbers of transactions—across these bases during FY 2004. A better understanding of the nature of activities at individual locations is necessary to draw conclusions; however, Figure 4.1 raises questions about the relative workload of CCOs at Balad. Their obligations and transactions were greater than those of CCOs at Kirkuk and Tallil, and they obligated approximately the same amount as CCOs at Baghdad but did it through far more transactions. In addition, our analyses of BPA purchases demonstrated that Baghdad utilized BPAs to a greater extent

Figure 4.1
Average Transactions and Dollars per CCO at Selected Locations, FY 2004

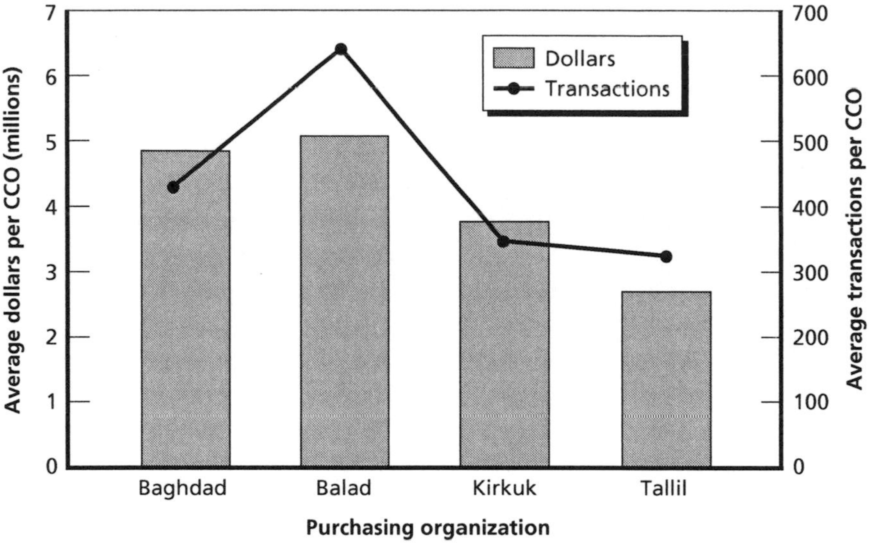

RAND *MG559/1-4.1*

than did Balad, which may indicate that, all else being equal, the fewer transactions per CCO at Baghdad also required less work for those CCOs.

By including the "missing" BQ dollars, we can compare average obligations per CCO at four additional organizations, as shown in Figure 4.2. Unfortunately, we cannot compare the number of transactions per CCO for these organizations.

We see even greater variation in obligations per CCO, with CCOs at Al Udeid and the CAOC obligating significantly more money than did their colleagues in other organizations. With additional information on the nature of the work within these organizations (e.g., mission activities supported, types of goods and services purchased, and the number of transactions completed), statistical analyses such as regressions could be used to understand the factors associated with these differences. However, these simple data displays offer motivation and tools for developing more systematic methods for assigning CCO personnel in theater.

Figure 4.2
Average Obligations per CCO, FY 2004 (including BQ data)

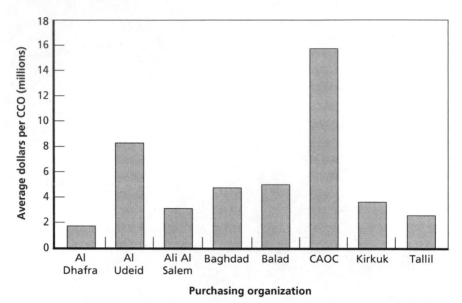

CCO Training

Anecdotes from our interviews indicate that a number of factors make contracting in theater challenging, including differences in the nature of contingency contracting duties as opposed to duties of a contracting officer at a nondeployed location, variation in the contracting environments among countries within the AOR, the short duration of most deployments for contracting personnel,[2] and differences in contracting culture among the military branches operating in a joint environment.[3]

---

[2] Typical deployments increased from three months to four months during our data time frame, FYs 2003 and 2004.

[3] During OIF, Air Force CCOs supported non–Air Force units and missions, and there were sometimes unexpected challenges in doing so. Some non–Air Force organizations manage

At first glance, there appears to be abundant guidance available to CCOs to help mitigate any adverse effects associated with these challenges. AFFARS Appendix CC "establishes policies, assigns responsibilities, and prescribes implementing procedures" for Air Force contingency contracting support (AFFARS, Appendix CC, para. CC-101). The organization and operation of Air Expeditionary Forces, including contracting functions, are described in Air Force Instruction (AFI) 10-401, Air Force Operations Planning and Execution (U.S. Department of the Air Force, 2005, Attachment 3). The Air Force Logistics Management Agency (AFLMA) also produces a contingency contracting handbook (Roloff, 2003).

In addition, formal training is available through DAU.[4] However, one officer we interviewed likened learning CCO procedures from such formal classroom training to learning to play golf by reading the rulebook. In contrast, several people we interviewed mentioned the importance of providing deploying CCOs with opportunities to engage in training simulations, e.g., Silver Flag exercises, that present them with scenarios they can expect to encounter when they go to the theater of operations.[5] In an attempt to provide "just-in-time," targeted training, CENTAF contracting offers predeployment orientation programs, but these are limited to office chiefs rather than all deploying personnel,

---

their contracting workforces in ways that differ from the Air Force's approach, entrusting less responsibility to enlisted contracting personnel and relying more heavily on officers for contracting activities. We were told that, as a result, highly skilled enlisted CCOs occasionally encountered resistance from officers in these other organizations who expected to be working with equivalent-rank officers. Because of different expectations, non–Air Force organizations would also make personnel requests by number of officers (e.g., three majors) rather than specifying needed capabilities that, in many cases, could be provided by Air Force enlisted CCOs.

[4] The course CON 234 (Contingency Contracting) is designed to help develop "skills for contracting support provided to Joint Forces across the full spectrum of military operations" (DAU, 2005, p. 36). At the time of this research, DAU was in the process of updating its contingency contracting curriculum.

[5] The Silver Flag exercises provide civil engineers, services, and other support personnel training on building and maintaining bare bases in deployed locations (GlobalSecurity.org, 2005a).

and, after completing the programs, personnel often return to their home bases for several weeks before deployment.

Finally, CCOs are required to maintain "continuity books" documenting their activities to help their replacements learn the status of contracts at that location, local suppliers and points of contact, and so forth. Although there is potential for this information to provide a rich context for understanding the types of raw data contained in the transaction databases we utilize in our analyses, we were told that there is a great deal of heterogeneity in the quality and completeness of the continuity books.

A database of CCO purchases such as the one we developed, which is described in Chapter Three, could supplement classroom and predeployment training by providing insights into ongoing activities in the theater. Information could be tailored to locations where trainees would be deploying. It could also assist in creating more realistic environments for exercises. In addition, a CCO who is getting ready to deploy could use the database to prepare by becoming familiar with the contracting environment at his or her future location, including the types of purchases made, the predominant types of contracts used for these purchases, and the local supply base. Similar data on contracting for other military branches and coalition partners could be used to better prepare CCOs who will be operating in a joint requirements environment.

Combat Support Planning

Combat support planners are responsible for making sure all the resources the Air Force needs to go to war are in place in time to support contingency operations and associated personnel. After determining all the necessary resources, planners must make choices about where to obtain them and how to get them to the theater to shorten the deployment-to-employment timeline, make the best use of scarce airlift and other transportation resources, and reduce the military footprint in theater. Since one option that planners consider is the availability of resources in theater, one of the motivations behind the development of this database was that such data could be used to improve combat support planners' ability to make effective, efficient trade-offs between purchasing items in theater and purchasing them elsewhere and then transporting them to the theater. The purchase of bottled water in Iraq provides a simple case study of how a detailed database of CCO purchases can be used to help assess the trade-offs among options, such as purchasing required assets as needed during operations in theater and purchasing them in advance and then using scarce airlift or other transportation assets to move materials from storage locations to operating locations. In addition, these data can be used to describe the local supply base for such purchases.

The U.S. military required a great deal of bottled water for personnel stationed in locations supporting OIF during FYs 2003 and 2004. Our database indicates that CCOs in 15 purchasing organizations in theater purchased bottled water through 38 contracts with more than 30 suppliers.[6] Alternatively, planners could have elected to set up contract vehicles for large quantities of water in advance (or purchase and store the water) and then ship the water to appropriate locations in theater as needed. Presumably, such advance planning would result in a lower cost per liter than CCOs were able to negotiate in real time during contingency operations; however, shipments of water into the theater would either delay the transport of troops and other supplies or would require the purchase of additional transportation.[7] A combat support planner can examine the costs of these options and any effects on mission to determine the best way to meet water requirements in theater during operations.

Our database can be used to assess the costs associated with purchasing water in theater and to assist with analysis of the amount of airlift required for an alternate approach. From January to March 2003, CCOs at seven purchasing organizations spent $402,993 on bottled water, with an average price per half-liter bottle of about 13 cents. From this information, planners can deduce demand in terms of

---

[6] The exact number of suppliers is uncertain due to ambiguities in supplier names.

[7] One or more contracts with regional providers that could easily distribute water to multiple locations would reduce the need for airlift.

the number of bottles. Supplemental information about water packaging and transportation capacity can then be used to evaluate alternatives. Table 4.2 shows some reasonable conclusions that can be drawn from this information.

Using the factors in Table 4.2, we can estimate transportation savings associated with purchasing water in theater. As an example, flying time from Ramstein AB in Germany (one supply base for the theater)

Table 4.2
Water Case Study: January 2003–March 2003

Characteristic	Amount or Quantity
Amount spent on water, January–March 2003	$402,993
Number of bottles purchased[a]	3.2 million
Weight of water purchased[b]	3.5 million pounds
Number of C-17–equivalent sorties required to transport this amount of water[c]	39
Number of troops that could be transported with equivalent airlift[d]	3,510
Cost of airlift required to ship an equivalent amount of water into theater	Approximately $5,000 per flying hour for C-17[e]

NOTE: Data are based on CCO bottled water transactions from January to March 2003 for the 24 purchasing organizations represented in our database. This example does not include water purchases that might be included in BQ data. Cargo and passenger capacity are from Table 3 in Air Force Pamphlet 10-1403 (U.S. Department of the Air Force, 2003, p. 12). The "planning" column in the table was used.

[a] Assumes that all water was purchased in 0.5-liter bottles with an average price of $0.126/bottle or $0.253/liter.

[b] One liter contains 33.8 fluid ounces of water and weighs 2.2 pounds.

[c] Cargo capacity is 90,000 pounds.

[d] One C-17 can carry 90 troops/paratroops.

[e] This cost estimate is based on the Air Force's standard CORE (Cost-Oriented Resource Evaluator) model, using input data from AFI 65-503, Table A54-1, updated December 10, 2004 (U.S. Department of the Air Force, 1994). It excludes personnel and other non–flying hour–related costs. We thank our RAND colleague Raymond A. Pyles for providing these data.

to Baghdad is about five hours,[8] so shipping the required amount of water from Ramstein into the theater would have cost approximately $975,000 (i.e., the cost of 39 one-way transport flights), in addition to the cost of purchasing the water. Given other demands for inter-theater airlift, purchasing water in theater freed up valuable air transport resources to move several thousand troops or military equipment. Table 4.2 does not include water purchases that were recorded in the BQ system, so the potential advantages of purchasing in theater may be underestimated.

In addition, data on joint contracting in theater, similar to those analyzed in this monograph, could be used by the combatant commands, e.g., CENTCOM, to construct more realistic and detailed contract support plans. These plans are intended to outline personnel requirements, organizational structures, and so forth, that will be used for joint contingency contracting to support operations executed by the combatant commands, e.g., at what point contracting should transition from a decentralized, service-specific structure to joint organizations.

Sharing Lessons

The nature of particular requirements and the local environment, including urgency and security threats, may limit CCOs' ability to reduce costs. However, awareness of details of purchases made by other CCOs in the theater should assist in negotiating better prices where this is possible. Table 4.3 shows the maximum, minimum, and average prices paid per liter of water in FYs 2003 and 2004 transactions in our database. (Note that the time period is different from the example above.)

---

[8] This five-hour flying time is mentioned in an article about the C-17's one-millionth flying hour. (See Lesser, 2006.)

Table 4.3
Range of Prices CCOs Paid per Liter of Drinking Water,
FYs 2003 and 2004

Category	Maximum	Minimum	Average
Price ($)	1.08	0.12	0.38
Date	March 2004	June 2003	
Location	Baghdad	Al Jaber	

NOTE: Prices are based on transactions for bottled water for which the price per liter was known under reasonable assumptions. For example, if "pallet" was the unit of purchase, it was assumed that the pallet contained 72 cases of 0.5-liter bottles, with 24 bottles per case. Because actual pallet and bottle sizes could be different, we excluded four "outlier" purchase prices obtained under these assumptions: low prices of $0.03/liter at Baghdad in May 2003 and $0.08/liter at Kirkuk in September 2004, and two high prices of $1.87/liter at Baghdad in March 2004 and $1.65 at Kirkuk in September 2004.

The purchase for Baghdad in Table 4.3 was for 64 pallets of bottled water, which, under our assumptions, equates to 110,592 half-liter bottles, or 55,296 liters. If the Baghdad CCO had been able to obtain this water for the price paid at Al Jaber, he or she would have saved more than $53,000. Of course, the majority of the cost for the Baghdad purchase may be attributable to the challenges of delivering into that location.

Table 4.4 shows the variation in water prices at locations across the region during FYs 2003 and 2004. Interestingly, Iraq had not only the highest prices, but also some of the lowest.

While price information can be a powerful tool for CCOs, additional information about the relative performance of suppliers and other factors related to meeting requirements, such as the urgency, transportation needs, or security threats, would be helpful in interpreting such comparisons.

Table 4.4
Prices Paid per Liter of Drinking Water in Theater, FYs 2003 and 2004

Location	High Amount ($)	High Year Paid	Low Amount ($)	Low Year Paid
Iraq				
Kirkuk	0.64	2003	0.14	2003
Baghdad	1.08	2004	0.16	2003
Tallil	0.51	2004	0.37	2004
Egypt				
Cairo West	0.29	2003	NA	NA
Kuwait				
Al Jaber	0.12	2003	NA	NA
Saudi Arabia				
Ar'Ar	0.38	2003	NA	NA
UAE				
Al Dhafra	0.25	2004	NA	NA
Al Minhad	0.34	2003	0.18	2003
Bahrain				
Shaikh Isa	0.29	2003	NA	NA
Oman				
Seeb	0.23	2003	0.22	2003
Masirah	0.27	2003	NA	NA

NOTE: NA = not available. There was insufficient information to estimate prices for four locations; thus this table includes only 11 of the 15 locations that purchased water. Some water purchases explicitly stated the number of bottles or liters purchased. Others stated that a certain number of pallets or cases of bottled water was purchased. As in Table 4.3, prices are based on reasonable assumptions (e.g., if "pallet" was the unit of purchase, it was assumed that the pallet contained 72 cases of 0.5-liter bottles, with 24 bottles per case—standard pallet and case sizes, according to a commercial vendor [Ultra Pure Bottled Water, Inc., undated]). The outliers noted in Table 4.3 are also excluded in Table 4.4.

CHAPTER FIVE

Recommendations

In this monograph, we have described the construction of a database of CCO purchases supporting Air Force activities in OIF during FYs 2003 and 2004. We have demonstrated how this database can be a powerful analytic tool to inform and support policy decisions and initiatives for contracting and combat support planning, programming, and budgeting decisions for forward bases, CCO assignments, and CCO training.

Based on our experience creating the database and analyzing the CCO data for OIF, we recommend that the Air Force (and DoD more broadly) establish a standardized methodology for collecting contingency contracting data on an ongoing basis to facilitate planning and policy decisions for future contingencies. Such a data collection effort would support policy decisions associated with CCO staffing and training, combat support planning, and sharing lessons within the theater.

When we began this study, there was essentially no standardization in the collection of CCO data; the spreadsheets collected by the CENTAF comptroller were the most uniform, but even they underwent some modification between FY 2003 and FY 2004. As a result, the data preparation needed to develop the analytic capability described in this monograph was extensive.

Since the beginning of our study, the Air Force has made progress toward developing more standardized databases and analytic capabilities. In FY 2003, CENTAF contracting introduced a PC-based data collection system called ACTT to standardize how CCOs tracked and provided information to CENTAF about completed transactions, dollars spent, and the number of CCO personnel at locations within the Persian Gulf region (Thrailkill, 2003). Although CENTAF developed ACTT in 2003, it was not widely or consistently used by CCOs until 2004.

ACTT represents a significant improvement over the heterogeneous PIIN logs formerly used by CCOs for reporting purposes. However, in its current form, it does not provide the level of detail about purchases found in the CENTAF comptroller's databases that we used for our analyses. For example, ACTT places purchases in the broad categories of construction, services, and commodities but does not require CCOs to provide more detailed information on the purchases.[1]

Similarly, the CENTAF comptroller introduced the Central Processing Site (CPS) system in November 2003. This system aggregates payment information by contract; however, to determine individual items that were purchased, one must obtain a copy of the contract and check the line items (see Griffith, c. 2005).

To facilitate the types of analyses illustrated here in a timely way, the Air Force needs to systematically gather contingency contracting data on an ongoing basis. To be most useful, the data system must make it possible to quickly access detailed descriptions of individual transactions, as well as aggregate those transactions according to categories of purchases, types of contract vehicles used, locations of purchases, suppliers dealt with, and so forth.

Table 5.1 contains our recommendations on the types of data that would be most useful to collect. These recommendations encompass data about the transactions themselves, as well as supplemental information about the activities supported by individual purchasing organizations and the relevant supply bases, that would enhance the types of analyses illustrated in this monograph and provide a basis for interpreting their results.

Creating some types of data is more time-intensive than is creating others. For example, for transactions covering multiple goods or services, we recommend disaggregating the total price paid to reflect

---

[1]   ACTT has a comments section for each transaction, but entries are not standardized.

Table 5.1
Recommended Data to Be Collected on an Ongoing Basis

Type of Data	Explanation
Individual transactions	Data to be entered by purchasing CCO
Purchasing organization	Organization that purchases the goods or services
CCO	Individual responsible for the transaction
Recipient	Organization or location that benefited from the purchase, if different from the purchasing organization, e.g., base that benefited from a RED HORSE repair project
Text description	Description of full range of goods and services purchased through the transaction
Units	Number of goods purchased or period of time for which service is to be provided; break out according to types of goods or services covered within the transaction
Purchase category	General class(es) of goods or services purchased; break out according to types of goods or services covered within the transaction
Price	Price paid for the goods and services; when multiple goods and services are purchased within a single transaction, prices should be broken out by type
Supplier	Firm that provides the goods and services
Location of supplier	Identifies whether supplier is a local firm, regional firm, or other
Transaction ID	Unique identifier for the transaction, e.g., contract number
Payment mechanism	GPC or contract
Type of contract	For contracts, type of contract, e.g., BPA, Form SF44
Date of request	Date on which purchasing organization received the formal request for goods and services
Date of payment	Date on which supplier was paid
Date of delivery	Date on which goods were delivered or services began
Comments	Any explanatory comments CCO deems useful

Table 5.1—Continued

Type of Data	Explanation
Activities supported by purchasing organizations	Supplemental data needed to explain purchasing trends; will vary over time
Population	Number of personnel supported by the purchasing organization
Mission activity	Description of mission activity supported by the purchasing organization's transactions, e.g., number and types of aircraft, special operations
Responsibility for BOS	Service branch responsible for providing BOS for the location
Infrastructure	Number of buildings, acres supported by the purchasing organization
Condition of infrastructure	Condition of infrastructure supported by the purchasing organization, particularly for new locations
Outlook	Plans for the purchasing organization, e.g., temporary operating location
Supply base	Supplemental data to facilitate improved purchasing over time
Supplier ratings	Performance ratings of suppliers (perhaps only key suppliers) based on, for example, the quality of goods and services, reliability, ease of working relationship

the prices paid for different types of goods or services. This would allow more opportunities to compare prices for similar commodity groups across the AOR. Similarly, *ex post* assessments of supplier performance can help with future source selection decisions and determining with which suppliers the Air Force may want to develop closer relationships.

We understand the complex and austere conditions in which CCOs often operate. Additionally, we do not propose to overburden these hard-working individuals with new reporting requirements. We do suggest a standardized automated system for transaction-specific data that could be either virtually connected to a master database or regularly downloaded into such a database as a means of recording and cataloging purchases. A system such as ACTT that includes the data elements found in the CENTAF comptroller logs that we used could satisfy both contracting and financial management requirements for information. Such a system should also include an easy method both for categorizing purchases across a wide range of commodities and services and for identifying suppliers in a standardized way. For example, drop-down menus with category options and supplier name options from which to choose would make it easier for CCOs to identify these in a consistent manner. Contingency contracting representatives and logistics planners should work in concert to develop the database, ensuring that one standardized system will satisfy the requirements of both organizations.

The Air Force is in the process of reviewing current contracting organizations, including those overseas, to determine what future organizations should look like. In addition, the Air Force is actively engaged in discussions about how to improve the effectiveness and efficiency of contracting in a joint contingency environment, in which forces from different military branches are collocated and are operating together. The analytic capabilities recommended in this monograph can provide key inputs to these important organizational and operational decisions.

Selected Contingency Contracting Issues

In this appendix, we discuss the general nature of battlefield contracting and the role of CCOs in supporting contingency operations.

Examples of Goods and Services Provided by Contractors

As noted in Chapter One, the Air Force has experienced growth in contractor support for contingency operations in recent years. Contractors provide a wide range of goods and services to deployed U.S. forces, freeing scarce military personnel for more important combat activities. While most goods are easy to imagine (anything ranging from major weapon systems to minor toiletry items that make base living more pleasant), the breadth of services is less fully appreciated. Here is a sample list of services provided by contractors at deployed locations (U.S. General Accounting Office, 2003, p. 7):

- weapon system support
- intelligence analysis
- linguists
- base operations support
- logistics support
- pre-positioned equipment maintenance
- nontactical communication
- generator maintenance
- biological/chemical detection systems
- management and control of government property

- command, control, communications, computers, and intelligence
- continuing education
- fuel and materiel transport
- security guards
- tactical and nontactical vehicle maintenance
- medical services
- mail services.

This list illustrates that contracted services do not simply provide straightforward support, such as military food service, but include complex services that directly affect or contribute to the missions of the U.S. armed forces.

Types of Contractor Support

According to Joint Publication 4-0, *Doctrine for Logistics Support of Joint Operations*, there are three broad categories of contractor support in theater: "systems support, external theater support, and theater support" (Joint Chiefs of Staff, 2000, p. V-1).

Doctrinally, *system support* contractors "provide logistics support to maintain and operate weapons and other systems," and contracts are typically awarded by the organizations responsible for procurement of the weapons or other systems (U.S. General Accounting Office, 2003, p. 6). Traditionally, this has been support for training and readiness on new, advanced, or recently fielded systems that the military has not yet developed the organic capability to support (Hammontree, 2003, p. 74). However, increasingly, system contractors are offering this support for the whole life cycle of the system, encompassing both peacetime and contingency use, in cases in which it is not cost-effective for the military to develop comparable capability organically. An example would be some of the advanced unmanned aerial vehicles (UAVs) deployed during OIF, which were maintained and operated solely by contractor personnel (Drew et al., 2005).[1]

External theater support contracts are preplanned support contracts awarded by organizations other than the component commands supporting the combatant commander (Hammontree, 2003; U.S. General Accounting Office, 2003). Examples include the Logistics Civil Augmentation Program (LOGCAP) and the Air Force Contract Augmentation Program (AFCAP) (Hammontree, 2003, p. 74).[2]

Theater support contractors provide goods and services to meet the immediate needs of the operational commander (Hammontree, 2003, p. 72). These contracts are "normally awarded by contracting agencies associated with the regional combatant command," such as component command headquarters or deployed bases (U.S. General Accounting Office, 2003, p. 5).

CCO Contracting

CCO contracting, which largely corresponds to "theater support contracting" in doctrine, is becoming a more intensive effort as more support activities are shifted to contractors. CCOs are often among the first forces to deploy into an area of operations and among the last to leave (Edwards, 2002).

---

[1] Prior to OIF, contractors supported Predator UAVs in Albania and Hungary as part of Operations Joint Promise in 1995 and Joint Endeavor in 1996 (CBO, 2005, p. 23).

[2] LOGCAP was established in 1985 as a preplanning tool for contractor support activities during contingencies or other crises and to utilize the private sector to supplement military forces (U.S. General Accounting Office, 2003, p. 6). Task orders for specific work are issued against an indefinite delivery/indefinite quantity contract that specifies a negotiated range of prices for a set of goods or services that can be provided by the contractor during a specific period. The first contract was awarded to Brown and Root Services in 1992. It was a five-year contract covering multiple locations, including the Balkans. The second five-year contract was awarded to DynCorp. The third (current) contract was awarded to Kellogg, Brown and Root in December 2001, with a 10-year performance period. See CBO (2005) for more details. AFCAP is a smaller logistics support contract. The eight-year cost-plus award fee contract was awarded to Readiness Management Support LC, a subsidiary of IAP Worldwide Services, in 2002 (CBO, 2005).

The CCO purchasing process begins with "users" (e.g., functional managers, the commander and his or her staff) who specify requirements that cannot be met within the normal supply system or through host-nation support, either at all or within the required time frame. If a financial management official can validate the requirement, it is the CCO's job to find a way to meet it in a timely, cost-effective manner.

The CCO's duties range from being one of the first personnel on the ground to help create (or "open") a new base to writing and maintaining contracts to support the needs of large groups of troops in the operating environment. In executing their duties, CCOs are often faced with challenges, such as finding appropriate sources while operating in remote locations and dealing with imposed constraints, such as requirements imposed by the host nation.

Opening a Base

CCOs are among the earliest arrivals when a new base needs to be opened. There may be little or no existing infrastructure, so a bare-base CCO is responsible for finding all the goods and services needed to create the base that are not part of the Basic Expeditionary Airfield Resources (BEAR) package or the regular supply system.[3]

> "In order to raise up a bare base, you need contracts," said Capt. Randy Culbreth, [then] commander of the 379th Expeditionary Contracting Squadron at Al Udeid AB, Qatar. "With Operation Enduring Freedom, most of the base services are contracted out. That means we go through local organizations, and not military assets for services." (Dospil, 2002)

At that specific base (Al Udeid), by late 2002, more than $46 million had been spent on more than 2,200 contracts, with approximately 30 percent of camp functions being contracted out (Dospil, 2002).

---

[3] The BEAR equipment consists of lightweight modular assets necessary to build a mobile air base and support deployed forces in an austere environment (GlobalSecurity.org, 2005b).

Finding Sources in Remote Locations

CCOs often operate in remote locations. Deployed bases may be established away from population centers, and, during recent contingencies, CCOs supported mobile units, such as special forces units, that sometimes operated in peripheral areas of the theater of operations.

Finding sources to meet requirements in remote locations (much less reliable, cost-effective sources) is inherently challenging. As Col Duane A. Jones, then chief of logistics for the Combined Forces Air Component Command, states, "'At first blush, you might ask why we'd deploy a contracting officer to an Iraqi air base early on, because where would we find vendors?' . . . The word got out, and the vendors came." Fortunately, as Col Jones observes, contracting officers do find local suppliers in out-of-the-way places. "Even contracting officers sent to remote areas found suppliers, some traveling great distances to do business with the coalition" (Chapman, 2003).

Host-Nation Issues

CCOs working in foreign countries may face constraints imposed by the host nations. Our interviews with CCOs suggest that supply base constraints are particularly challenging. Host nations may limit the effective supply base that CCOs can draw on through direct means, requiring special licensing for firms to do business with the U.S. military, or through indirect means such as imposing substantial delays in clearing customs. Our supplier analyses demonstrate that CCOs can and do purchase needed goods and services from U.S. firms. Goods can be delivered directly to a base via military airlift, avoiding customs; however, this diverts transportation resources from other military requirements. In some cases, commercial carriers are able to fly directly onto bases during contingency operations; however, this is not without risk. For example, in November 2003, a DHL aircraft was severely damaged by a surface-to-air missile as it left Baghdad International Airport (see CNN, 2003).

Data Sources for Purchases Supporting Air Force Missions in OIF

We began this research by seeking sources for data on purchases supporting Air Force activities in OIF. Through discussions with the research sponsor, contracting personnel at Headquarters, Air Combat Command (ACC), and contracting personnel at CENTAF, we constructed a list of potential data sources. We were able to collect relevant data from virtually all of the suggested sources, though comprehensive data were not available in all cases. Table B.1 summarizes the data we collected from relevant organizations.

Table B.1
Data Sources and Available Relevant Data

Data Source	Relevant Data	Actions Taken
CENTAF	CCO purchases at OIF bases in the CENTCOM AOR	Met with contracting and financial management personnel at 9th Air Force and Al Udeid AB, Qatar Collected all available data See Tables 2.2 and 3.1
USAFE	CCO purchases at OIF bases in the U.S. European Command AOR	Met with contracting personnel at Ramstein AB Collected available CCO data $25.5 million and 980 transactions between mid-February and early May 2003
Pacific Air Forces	CCO OIF purchases within the U.S. Pacific Command AOR	Collected available CCO data (GPC only) $129,000 and 97 transactions between mid-March and early June 2003

Table B.1—Continued

Data Source	Relevant Data	Actions Taken
AFSOC	CCO OIF activities associated with AFSOC units and missions	Met with contracting personnel at Hurlburt AFB Collected available CCO data. Data are likely incomplete, and it was difficult to disentangle OIF from other types of expenditures $150,000 and 120 transactions between April 2003 and April 2004
CENTAF	Functional purchases for Air Force bases in the CENTCOM AOR	Met with functional personnel at 9th Air Force and WRM personnel at Al Udeid AB, Qatar Collected data from civil engineers and communication, logistics, force protection, operations, and intelligence personnel See Table 1.1
Air Force Civil Engineer Support Agency	AFCAP purchases for theater logistics support	Met with program manager at Tyndall AFB Collected more detailed data on all relevant actions, but missing some location data; difficult to separate locations supporting OIF from other locations within the AOR Data supplement the civil engineer data collected from CENTAF
AMC	CCO purchases associated with tanker airlift control element (TALCE) operations	Received all existing CCO data for TALCE, but they are incomplete $3.7 million and almost 700 transactions between December 2002 and July 2003
AMC	Contract airlift	Received some aggregate contract airlift data, but it is not possible to separate marginal expenditures associated with OIF from the cost of airlift services supporting other Air Force activities around the world
U.S. Transportation Command	PowerTrack transportation purchases	None to date
Defense Energy Support Center	Fuels	None to date
U.S. Space Command or Defense Information Systems Agency	Satellite capacity, bandwidth	None to date

Preparing the CCO Data for Analyses

In this appendix, we describe four processes we used to transform the original CCO data files provided by the CENTAF comptroller into data that could be used for our analyses. These processes included correcting errors in the data, creating new variables, grouping purchases into categories, and standardizing supplier names.

Data Cleaning

Few databases are free of errors. However, identifying and correcting errors can be time-consuming, and researchers must be careful to weigh the value added against the resources required to correct the data. Two variables in the CENTAF comptroller databases that were important to our analyses contained errors that were relatively easy to identify and correct: the date of the request and the date the supplier was paid.

We used two processes to identify date errors. First, we looked for transactions whose dates paid were listed as occurring prior to the dates of the requests, suggesting that a purchase took place in advance of a stated need. We then examined transactions listed before and after each of these to identify and correct the variable in error. Second, we looked for transactions whose dates of request occurred outside the fiscal year of the relevant database, e.g., a November 2003 date of request for a transaction found in an FY 2003 file. We looked at transactions listed before and after each of these, as well as the date paid, to determine whether the date was really in error. Often, these types of problems were associated with misuse of the fiscal year rather than the calendar year (e.g., an FY 2003 transaction occurring in November 2002 was listed as occurring in November 2003).

Creation of New Variables

To simplify our analyses, we created a variable for the purchasing organization (e.g., Baghdad, CAOC) associated with each transaction. We used the name of the original data file to create this variable.

We also created a fiscal year variable based on the date of request for each transaction.[1] Where this information was missing, we used the date paid as a proxy.

Finally, we created a transaction type variable describing the payment mechanism (i.e., GPC or contract, type of contract) used for each transaction. We could not identify the type of contract in all cases; however, we were able to consistently identify BPA transactions. Therefore, this variable takes on three values: GPC, BPA, and other contract. We used three processes to identify the type of contract. Sometimes the original data spreadsheet contained a data field identifying whether a contract transaction occurred via BPA; however, this data field did not identify all BPA transactions. We used the text description, which often contained the term "BPA," to supplement this information. Finally, we used the contract number to identify the comprehensive set of transactions (referred to as *calls*) against an identified BPA.

Categorization Difficulties

As discussed in Chapter Three, we grouped purchased goods and services into 45 categories for our analyses (see Table C.1). These categories were chosen based on our examination of the data and were

---

[1]   Data contained in original files from the CENTAF comptroller sometimes fell outside the fiscal year boundary.

validated through our interactions with Air Force CCOs.[2] While many categories are fairly broad, e.g., construction supplies, some narrow categories were used for classes of purchases that were easy to define and identify, e.g., bottled water was kept separate from food. Having a separate category for these does not preclude their being combined for types of analyses that differ from those illustrated in this monograph. Some categories, such as MWR or force protection, are based on usage rather than on any class of commodity.

We developed a computer program to ensure consistency in the categorization process. The program uses key words or phrases from transaction descriptions to determine which category or categories best describe a transaction. For example, if a transaction description includes the word *concrete* along with the word *barrier,* the transaction would be included in the force protection category. Though time-consuming to create, this program allows us to easily shift items from one category to another as appropriate.

The drawback of such a categorization process is that it does not lend itself to accommodating nuances in descriptions of purchases. For example, *workstation* is used to describe both office furniture and computers. Often, one can easily tell which is correct by the context of the rest of the purchase description; however, it is difficult to create a computer program to do this. Therefore, in consultation with our research sponsor, we sought to achieve 90-percent accuracy in our categorizations, and we validated this level of accuracy by examining purchases selected for each of the categories.[3] Table C.1 lists each purchase cat-

---

[2] Standard classification schemes such as the North American Industry Classification System (NAICS) seemed overly complex and, in some cases, insufficiently precise for the types of analyses described in this monograph. For example, while there is a distinct code for janitorial services (561720), the most relevant code for bottled water is 454390 (Other Direct Selling Establishments), which includes such industries as fruit stands and Christmas tree farms in addition to providers of bottled water. (See U.S. Census Bureau, 2002.)

[3] Members of the research team examined the transactions selected for each category. We focused on transactions selected only for that category, e.g., only computer equipment, as well as those selected for an additional category, e.g., computer equipment and office supplies.

egory used in this research. The table provides examples of purchases included in each category.

Category assignments via key words and phrases are sometimes complicated by the wide variety of potential ways to describe individual purchases. For example, the portable memory device that can be plugged into a USB port on personal computers was described by different CCOs as a "USB drive," "USB storage," "flash drive," "flash memory," "memory stick," "mem stick," "micro vault," "thumbdrive," "thumb drive," and more. While checking the accuracy of the computer assignments, we had to ensure that our program accounted for each of the relevant descriptions.

Another challenge we encountered was the originality of spelling and abbreviations observed in transaction records. For example, different CCOs described refrigerator purchases as a "fridge," "refrig," "refrigeration," "refridgerator," and even "regfrigerator." Similarly, a popcorn popper purchase was described as a "popcorn pooper." The nature of our categorization program requires us to associate each of these variations with particular categories. Although we do not expect CCOs to check the spelling of each of their transaction entries, these variations did make our categorization process more difficult and time-consuming.

Finally, some purchase descriptions do not contain enough detail to allow us to assign them to categories. For instance, descriptions such as "2 Sch 80 90 Deg Sweep" and "SF44 1 275 70 R22 8" had to be assigned to an "unknown" category.

Standardization of coding options for purchase descriptions, as discussed in our recommendations, would help eliminate these sorts of difficulties.

Table C.1
Categories Used in Our Analyses

Category	Examples
Appliances	Laundry (washers and dryers) Kitchen (refrigerators, kitchen ranges, microwave ovens, dishwashers) Miscellaneous (water heaters, air conditioners, ceramic heaters, ice machines)
Billeting services	Billeting (apartment rental, leasing of rooms) Downtown stays (hotel lodging, room bills)
Buildings and shelters	Residential buildings (living quarters, trailers) Structures (clamshell buildings, dome structures, prefabricated buildings, modular units, gazebos, bedouin tents) Facilities (storage buildings, shower trailers, field showers, water-treatment plants)
Cleaning supplies	Cleansers (detergents, dishwashing liquid, laundry soap, glass cleaner) Cleaning supplies (rags, brushes, rubber gloves, brooms, mops)
Communication equipment	Local area network equipment (server, high-speed network equipment, Ethernet catalyst switches [Ethernet equipment other than cards], coaxial cable, data cable, Cisco® switches, fiberoptic items, routers, Linksys® boxes, X-port switches, Secret Internet Protocol Router Network [SIPRNET] equipment) Communication systems (news dishes, uninterruptible power supply systems, videoconference equipment) Personal devices (radio equipment, handsets)
Computer equipment and software	Computers (desktops, laptops, keyboards, mice, computer monitors, computer speakers) Computer drives (hard drives, memory sticks) Computer accessories (personal digital assistants, scanners, CD burners, DVD burners, computer power supply) Server connections (USB hubs and cables, Ethernet cards, modems) Software (Adobe® Acrobat®, Microsoft® Windows® licenses)
Construction, heavy equipment	Backhoes, loaders, bulldozers, dump trucks, excavators, graders, trenchers
Construction services	Preparation (soil stabilization, clearing, digging, soil surveys) Building (construction work, road construction, ramp construction) Clearing (demolition/teardown, tree removal) Miscellaneous (airfield marking, sandbag services, various renovations and upgrades, installation of equipment, connect/install generators)

Table C.1—Continued

Category	Examples
Construction supplies	Hardware (nails, screws, nuts, bolts, washers) Construction material (steel, concrete, cement, asphalt, wood, plywood, sand, rock, gravel, 2x4s, planks, crossbeams) Electrical material (circuit boards, grounding material, cable) Plumbing material (pipe, toilets) Finishing material (carpet, floor covering, tile, sealant, stains, paint, painting equipment, bathroom fixtures) Runway construction and repair material Miscellaneous (ladders; culverts; manhole covers; heating, ventilation, and air conditioning)
Custodial and latrine services	Cleaning (latrine trailers, hangars) Custodial services Janitorial services
Dining supplies	Cooking utensils (spatulas, spaghetti tongs, can openers, cooking thermometers) Kitchen supplies (coffee pots, mixers, canisters, pans, aluminum foil, salt and pepper shakers) Serving supplies (dining trays, paper products, plastic utensils, food containers) Large equipment (pastry cases, beverage dispensers, salad bars) Other (aprons, tablecloths)
Financial	Fees (account maintenance fees, transaction charges, currency exchange, electronic funds transfer fees) Checkbooks Rebates (International Merchant Purchase Authorization Card [IMPAC]/GPC rebates)
Fire protection	Equipment (fire extinguishers, fire bottles, flame-retardant hoods, smoke alarms, smoke detectors, fire helmets, firefighter equipment) Training materials
Food (not catering)	Food (bread, cake, popcorn) Drinks (sports beverages) Cooking ingredients (cooking oil, salt)
Food service	Catering

Table C.1—Continued

Category	Examples
Force protection	Barricades (concrete barriers, roadblock spikes, barbed wire, concertina wire, chain-link fencing, cones, sandbags) Dog-related equipment (kennels, food, supplies) Surveillance (motion detector, walk-through metal detectors, gas detectors, search pit equipment, guard towers, metal detectors, floodlights) Personal and vehicle IDs Miscellaneous (badge-activated locks, reflective belts, reflective tape, biodetection/protection equipment) Police-related items (lightbars, blood-alcohol detection meters, handcuffs)
Fuel and fuel-related items (not jet fuel)	Fuels (diesel, acetylene, propane) Fuel-storage equipment (fuel tanks, fuel bladders) Fuel-dispensing equipment
Furniture	Office (desks, chairs, couches, bookcases, filing cabinets, workstations) Residential (beds, mattresses, dressers, footlockers) Other (stools, rugs, seats, cabinets, tables, folding chairs, paintings)
Generators	Various power generators
Grounds maintenance services	Groundskeeping services
Heavy equipment (not construction)	Large vehicles (refrigerated trucks, firetrucks, flatbed trucks, sewage-removal trucks, water trucks, fuel trucks, freezer trucks) Cranes, forklifts, bucket loaders, aircraft stairways
Interpreter services	Interpreters, linguists, and translator services
Latrine supplies	Shower and bathroom supplies (soap, waterless hand cleanser, paper towels) Chemicals for portable toilets
Laundry services	Laundry and dry cleaning Linen exchange Alterations and embroidery Self-serve laundry centers
Medical services	Doctor, dental, optometry, and chiropractic services Hospital charges Magnetic resonance imaging, X-ray consultation Biohazard disposal

Table C.1—Continued

Category	Examples
Medical supplies	Medical supplies (bandages, thermometers, sterile water, medication, insulin, vaccines, syringes) Medical equipment (X-ray equipment, dental equipment, respirators, lab equipment, monitors) Medical reference books Mortuary-affairs items
Miscellaneous commodities	Items for personnel (T-shirts for various activities [not MWR, not mission], backpacks, gloves, knives, towels, duffel bags, irons, duct tape, keys, bed linens, window treatments, baby wipes, sunscreen) Nonpotable water (bulk water, dry ice) Small containers (hard-sided cases) Small equipment (locks, coolers/ice chests, small heaters, scales, batteries [not for cars], cigarette butt cans, cameras, video recorders, ear protectors, flashlights, irons, voltage converters/adapters, absorbent mats, air filters) Other miscellaneous items (insect bait, weed killer, mousetraps, flags, etiquette books, signs, antifatigue mats, spill kits, lamps, mirrors [not specific to other categories], filters [generic], wastepaper baskets)
Miscellaneous equipment	Small equipment (mortar mixer, wet and dry vacuums, pumps, refrigeration units, air compressors, blowers, hedge trimmers, Coleman® products, portable vacuums, fans, plasma monitors [not TVs]) Large containers (shipping containers, tanks, food and trash containers, steel drums, intermodal containers) Food/water screening (water-detection equipment, salmonella screening kits) Hard-to-categorize items (cash counters, bullhorns, megaphones, hand-washing stations, photo lab accessories, turbidimeters, pallets, trolley jacks, locksmith equipment, adapters [not specific to other categories])
Miscellaneous services	Miscellaneous (vehicle registration and licensing, photo developing, locksmith services, Internet services, picking up litter, photocopying, engraving, storage handling, airfield sweeping, grease removal [including cleaning grease traps]) Professional services (consultant services)

Table C.1—Continued

Category	Examples
Mission equipment	For personnel (flying glasses, portable personal latrine systems, armor, wicking undergarments, cold-weather gear, protective gear, assault backpacks, hydration packs, laser sights, rangefinders, aviator oxygen, gun supplies, holsters, M-16 equipment, bomb suits) Equipment (aircraft service carts, air traffic training equipment, special operations kits, hazmat containers, rescue equipment, training obstacles, explosive ordnance disposal training equipment, honor guard equipment, safes for storage, foreign-object debris containers, global positioning systems, flammable materials storage, ammunition storage) Miscellaneous (Arabic dictionaries)
MWR	Fitness center goods and services (weights, bicycles, fitness equipment, sports equipment, fitness center maintenance) Pool-related goods and services (pool supplies, pool cleaning equipment, pool maintenance) Entertainment (TVs, satellite TV service, high-definition TV, cable TV, computer games, CD and DVD players, outdoor grills, books, movies, CDs, party supplies, popcorn machines, stereos, grilling supplies, mobile pedestals) Holiday decorations Religious items (books, chapel items, other worship-related items) Morale awards (T-shirts for special programs, certificates, coins) Other (newspapers, newsletters)
Office supplies and equipment	Paper products (paper, files, folders, card stock, boxes, stationery) Administrative supplies (staples, staplers, hole punches, printer supplies, desk sets, Form SF44 books, ID holders) References (Federal Acquisition Regulation, Defense Federal Acquisition Regulation) Mailing supplies (stamps, tape, labels, bubble wrap) Computer storage devices (CDs, DVDs, floppy disks) Copiers and printers, plotters, fax machines Miscellaneous (laser pointers, shredding machines, white boards, canned air dusters, typewriters)
Other repair parts (not for vehicles)	Test equipment (test kits, circuit testers, detectors, meters, oscilloscopes, gauges) Replacement parts (bench stock, filters [if not categorized elsewhere], fluids, sealants, refrigerants)
Phones and services	Mobile phones, cell phones, satellite phones, related services Single in-line memory module (SIMM) cards Telephone chargers

Table C.1—Continued

Category	Examples
Refuse and garbage services	Refuse and garbage services Trash/waste collection and removal
Repair/maintenance services	Service contracts Item repair and maintenance (bicycles, vehicles, generators, motors, engines, alternators, pumps) Calibration
Sewage and portable latrine services	Septic and sewage services Portable latrine cleaning services
Tools	Basic tools (hammers, screwdrivers, drills, drill bits, clamps) Other tools (multipurpose tools, pressure sprayers) Welding and soldering equipment
Transporting cargo	Express mail fees and other shipping charges, delivery charges Customs fees
Transporting people	Airfare Emergency leave Taxi and limousine charges
Uniforms	Honor guard T-shirts, military boots, brassards Insignias and patches (enlisted rank, CENTAF patches, desert patches)
Utility services	Electricity charges
Vehicle repair parts	Equipment (tow vehicle equipment, battery chargers) Parts (tires, radiators, starters, belts, clutches, shock absorbers, radiator hoses, wiper blades, oil filters, pumps, switches) Fluids (transmission fluid, motor oil)
Vehicles for transportation	Passenger vehicles (autos, buses, sedans, light trucks, sport-utility vehicles) Other small vehicles (pickup trucks, all-terrain vehicles, John Deere® Gator™ utility vehicles)
Water	Potable water Potable ice

Standardizing Supplier Names

As with descriptions of purchases, there are often multiple ways to identify the firm providing a good or service. As an example, Grainger was also described as "Grainger Com," "Grainger Website," "Graingers,"

and "Granger." To accurately identify the top suppliers, we needed
to consolidate purchases associated with a single firm as accurately as
possible; however, the enormous number of supplier names made it
infeasible for us to consolidate purchases for every supplier. Instead,
we used the original data to identify the top 200 suppliers, according
to obligated dollars. We examined the entire list of supplier names to
identify alternative spellings for each of those 200 suppliers and then
consolidated the purchases associated with those firms. As our analyses
in Chapter Three indicated, the top 10 firms accounted for more than
40 percent of the dollars in our CCO database. Therefore, we feel con-
fident that we have correctly captured purchases associated with the
most important suppliers for the locations of interest to our study.

References

AFAA—*see* Air Force Audit Agency.

AFFARS—*see* Air Force Federal Acquisition Regulation Supplement.

Air Force Audit Agency, *Installation Report of Audit: Blanket Purchase Agreements 379th Air Expeditionary Wing Al Udeid AB, Qatar*, F2004-0057-FDE000, June 22, 2004.

Air Force Federal Acquisition Regulation Supplement, Contracting Laboratory, Hill AFB, Utah. As of May 9, 2006:
http://farsite.hill.af.mil/

Amouzegar, Mahyar A., Robert S. Tripp, Ronald G. McGarvey, Edward W. Chan, and Charles Robert Roll, Jr., *Supporting Air and Space Expeditionary Forces: Analysis of Combat Support Basing Options*, Santa Monica, Calif.: RAND Corporation, MG 261-AF, 2004. As of October 13, 2006:
http://www.rand.org/pubs/monographs/MG261/

Ausink, John A., Laura H. Baldwin, Sarah Hunter, and Chad Shirley, *Implementing Performance-Based Services Acquisition (PBSA): Perspectives from an Air Logistics Center and a Product Center*, Santa Monica, Calif.: RAND Corporation, DB-388-AF, 2002. As of October 13, 2006:
http://www.rand.org/pubs/documented_briefings/DB388/

Ausink, John A., Laura H. Baldwin, and Christopher Paul, *Air Force Procurement Workforce Transformation: Lessons from the Commercial Sector*, Santa Monica, Calif.: RAND Corporation, MG-214-AF, 2004. As of October 13, 2006:
http://www.rand.org/pubs/monographs/MG214/

Ausink, John A., Frank Camm, and Charles Cannon, *Performance-Based Contracting in the Air Force: A Report on Experiences in the Field*, Santa Monica, Calif.: RAND Corporation, DB-342-AF, 2001. As of October 13, 2006:
http://www.rand.org/pubs/documented_briefings/DB342/

Baldwin, Laura H., John A. Ausink, Nancy F. Campbell, John G. Drew, and Charles Robert Roll, Jr., *Analyzing Contingency Contracting Purchases for Operation Iraqi Freedom*, Santa Monica, Calif.: RAND Corporation, 2008. Not available to the general public.

Baldwin, Laura H., John A. Ausink, and Nancy Nicosia, *Air Force Service Procurement: Approaches for Measurement and Management*, Santa Monica, Calif.: RAND Corporation, MG-299-AF, 2005. As of October 13, 2006:
http://www.rand.org/pubs/monographs/MG299/

Baldwin, Laura H., Frank Camm, Edward G. Keating, and Ellen M. Pint, *Incentives to Undertake Sourcing Studies in the Air Force*, Santa Monica, Calif.: RAND Corporation, DB-240-AF, 1998. As of October 13, 2006:
http://www.rand.org/pubs/documented_briefings/DB240/

Baldwin, Laura H., Frank Camm, and Nancy Y. Moore, *Strategic Sourcing: Measuring and Managing Performance*, Santa Monica, Calif.: RAND Corporation, DB-287-AF, 2000. As of October 13, 2006:
http://www.rand.org/pubs/documented_briefings/DB287/

Cahlink, George A., "Send in the Contractors," *Air Force Magazine*, Vol. 86, No. 1, January 2003. As of October 16, 2006:
http://www.afa.org/magazine/jan2003/0103contract.asp

Calbreath, Dean, "MZM Scandal Illuminates Defense Contract Tactics," *San Diego Union-Tribune*, August 21, 2005. As of October 16, 2006:
http://www.signonsandiego.com/news/politics/cunningham/
20050821-87-mzmscand.html

Camm, Frank, and Victoria A. Greenfield, *How Should the Army Use Contractors on the Battlefield? Assessing Comparative Risk in Sourcing Decisions*, Santa Monica, Calif.: RAND Corporation, MG-296-A, 2005. As of October 16, 2006:
http://www.rand.org/pubs/monographs/MG296/

CBO—*see* Congressional Budget Office.

Chapman, Suzann, "Ability to Buy Local a Boon," *Air Force Magazine*, Vol. 86, No. 6, June 2003, p. 17. As of October 16, 2006:
http://www.afa.org/magazine/June2003/0603world.pdf

CNN, "Suicide Car Bombings Kill at Least 18 Iraqis, Cargo Plane Leaving Baghdad Airport Damaged by SAM," November 22, 2003. As of October 16, 2006:
http://www.cnn.com/2003/WORLD/meast/11/22/sprj.irq.main/

Congressional Budget Office, *Logistics Support for Deployed Military Forces*, Washington, D.C., October 2005. As of October 16, 2006:
http://www.cbo.gov/ftpdocs/67xx/doc6794/10-20-MilitaryLogisticsSupport.pdf

DAU—*see* Defense Acquisition University.

Defense Acquisition University, *2006 Defense Acquisition University Catalog*, Ft. Belvoir, Va.: DAU Press, October 2005.

Defense Finance and Accounting Service, *Standard Base Level General Accounting and Finance System User Manual*, DFAS DE 7077.2-M, Indianapolis, Ind., 2000.

———, *Defense Finance and Accounting Service Manual 7097.01 for FY 2004*, department reporting manual for Office of the Secretary of Defense (Treasury Index 97) Appropriations, Indianapolis, Ind., September 2003.

DFAS—*see* Defense Finance and Accounting Service.

Dixon, Lloyd, Chad Shirley, Laura H. Baldwin, John A. Ausink, and Nancy F. Campbell, *An Assessment of Air Force Data on Contract Expenditures*, Santa Monica, Calif.: RAND Corporation, MG-274-AF, 2005. As of October 13, 2006: http://www.rand.org/pubs/monographs/MG274/

DoD—*see* U.S. Department of Defense.

Dospil, Tarkan, "Contracts Keep Unit Operating in War," press release, U.S. Air Force, July 30, 2002. As of October 16, 2006: http://www.findarticles.com/p/articles/mi_prfr/is_200207/ai_2868902561/print

Drew, John G., Russell D. Shaver, Kristin F. Lynch, Mahyar A. Amouzegar, and Don Snyder, *Unmanned Aerial Vehicle End-to-End Support Considerations*, Santa Monica, Calif.: RAND Corporation, MG-350-AF, 2005. As of October 13, 2006: http://www.rand.org/pubs/monographs/MG350/

Edwards, Travis, "Contingency Contracting—Logistics Force Multipliers," Army News Service, October 29, 2002. As of May 19, 2006: http://www.globalsecurity.org/military/library/news/2002/10/mil-021029-usa01.htm

Galway, Lionel A., Robert S. Tripp, Timothy Ramey, and John G. Drew, *Supporting Expeditionary Aerospace Forces: New Agile Combat Support Postures*, Santa Monica, Calif.: RAND Corporation, MR-1075-AF, 2000. As of October 13, 2006: http://www.rand.org/pubs/monograph_reports/MR1075/

GlobalSecurity.org, "Silver Flag," last updated August 21, 2005a. As of October 16, 2006: http://www.globalsecurity.org/military/ops/silver-flag.htm

———, "Basic Expeditionary Airfield Resources (BEAR)," last updated September 7, 2005b. As of October 16, 2006: http://www.globalsecurity.org/military/systems/aircraft/systems/bear.htm

Griffith, Capt Michelle L., "Central Processing Site (CPS)," U.S. Central Command Air Forces briefing, c. February 10, 2005.

Hammontree, George (Sam) III, "Contractors on the Battlefield," *Logistics Dimensions*, Maxwell AFB, Ala.: Air Force Logistics Management Agency, 2003, pp. 69–84.

Joint Chiefs of Staff, *Doctrine for Logistics Support of Joint Operations*, Joint Publication 4-0, April 6, 2000. As of October 16, 2006: http://www.dtic.mil/doctrine/jel/new_pubs/jp4_0.pdf

Leftwich, James, Robert Tripp, Amanda Geller, Patrick Mills, Tom LaTourrette, C. Robert Roll, Jr., Cauley von Hoffman, and David Johansen, *Supporting Expeditionary Aerospace Forces: An Operational Architecture for Combat Support Execution Planning and Control*, Santa Monica, Calif.: RAND Corporation, MR-1536-AF, 2003. As of October 16, 2006: http://www.rand.org/pubs/monograph_reports/MR1536/

Lesser, Gary, "One in a Million," *Boeing Frontiers*, Vol. 5, No. 1, May 2006. As of September 8, 2006: http://www.boeing.com/news/frontiers/archive/2006/may/cover1.html

Lynch, Kristin F., John G. Drew, Robert S. Tripp, and Charles Robert Roll, Jr., *Supporting Air and Space Expeditionary Forces: Lessons from Operation Iraqi Freedom*, Santa Monica, Calif.: RAND Corporation, MG-193-AF, 2005. As of October 16, 2006: http://www.rand.org/pubs/monographs/MG193/

Moore, Nancy Y., Laura H. Baldwin, Frank Camm, and Cynthia R. Cook, *Implementing Best Purchasing and Supply Management Practices: Lessons from Innovative Commercial Firms*, Santa Monica, Calif.: RAND Corporation, DB-334-AF, 2002. As of October 16, 2006: http://www.rand.org/pubs/documented_briefings/DB334/

Moore, Nancy Y., Cynthia R. Cook, Clifford A. Grammich, and Charles Lindenblatt, *Using a Spend Analysis to Help Identify Prospective Air Force Purchasing and Supply Management Initiatives: Summary of Selected Findings*, Santa Monica, Calif.: RAND Corporation, DB-434-AF, 2004. As of October 13, 2006: http://www.rand.org/pubs/documented_briefings/DB434/

Moseley, Lt Gen T. Michael, *Operation Iraqi Freedom—By the Numbers*, U.S. Central Command, Air Forces, Assessment and Analysis Division, April 30, 2003. As of October 16, 2006: http://www.globalsecurity.org/military/library/report/2003/uscentaf_oif_report_30apr2003.pdf

Pint, Ellen M., and Laura H. Baldwin, *Strategic Sourcing: Theory and Evidence from Economics and Business Management*, Santa Monica, Calif.: RAND Corporation, MR-865-AF, 1997. As of October 13, 2006: http://www.rand.org/pubs/monograph_reports/MR865/

Robins AFB and Warner Robins Air Logistics Center, *Fiscal Year 01 Annual History*, Robins AFB, Ga., 2001.

Roloff, James, *Contingency Contracting: A Handbook for the Air Force CCO*, Maxwell AFB, Ala.: Air Force Logistics Management Agency, February 2003. As of October 16, 2006:
http://www.aflma.hq.af.mil/lgj/contingency%20Contracting%20Mar03_corrections.pdf

Snyder, Don, and Patrick Mills, *Supporting Air and Space Expeditionary Forces: A Methodology for Determining Air Force Deployment Requirements*, Santa Monica, Calif.: RAND Corporation, MG-176-AF, 2004. As of October 13, 2006:
http://www.rand.org/pubs/monographs/MG176/

Thrailkill, Capt Doug, "USCENTAF Automated Contract Tracking Tool (ACTT)," briefing, September 2003.

Tripp, Robert S., Lionel A. Galway, Paul Killingsworth, Eric Peltz, Timothy Ramey, and John G. Drew, *Supporting Expeditionary Aerospace Forces: An Integrated Strategic Agile Combat Support Planning Framework*, Santa Monica, Calif.: RAND Corporation, MR-1056-AF, 1999. As of October 13, 2006:
http://www.rand.org/pubs/monograph_reports/MR1056/

Tripp, Robert S., Lionel A. Galway, Timothy Ramey, Mahyar A. Amouzegar, and Eric Peltz, *Supporting Expeditionary Aerospace Forces: A Concept for Evolving the Agile Combat Support/Mobility System of the Future*, Santa Monica, Calif.: RAND Corporation, MR-1179-AF, 2000. As of October 13, 2006:
http://www.rand.org/pubs/monograph_reports/MR1179/

Tripp, Robert S., Kristin F. Lynch, John G. Drew, and Edward W. Chan, *Supporting Air and Space Expeditionary Forces: Lessons from Operation Enduring Freedom*, Santa Monica, Calif.: RAND Corporation, MR-1819-AF, 2004. As of October 16, 2006:
http://www.rand.org/pubs/monograph_reports/MR1819/

Tripp, Robert S., Kristin F. Lynch, Ronald G. McGarvey, Don Snyder, Raymond A. Pyles, William A. Williams, and Charles Robert Roll, Jr., *Strategic Analysis of Air National Guard Combat Support and Reachback Functions*, Santa Monica, Calif.: RAND Corporation, MG-375-AF, 2006. As of October 13, 2006:
http://www.rand.org/pubs/monographs/MG375/

Tripp, Robert S., Kristin F. Lynch, Charles Robert Roll, Jr., John G. Drew, and Patrick Mills, *A Framework for Enhancing Airlift Planning and Execution Capabilities Within the Joint Expeditionary Movement System*, Santa Monica, Calif.: RAND Corporation, MG-377-AF, 2006. As of October 13, 2006:
http://www.rand.org/pubs/monographs/MG377/

Ultra Pure Bottled Water, Inc., "FAQs," undated Web page. As of October 16, 2006:
http://www.nameyourbottle.com/faq.htm

U.S. Census Bureau, *North American Industry Classification System, Revisions for 2002*, Washington, D.C., 2002. As of August 29, 2006:
http://www.census.gov/epcd/naics02/

U.S. Department of the Air Force, U.S. Air Force Cost and Planning Factors, Air Force Instruction 65-503, February 4, 1994. As of September 20, 2007:
http://www.e-publishing.af.mil/shared/media/epubs/AFI65-503.pdf

———, "Air Mobility Planning Factors," Air Force pamphlet 10-1403, December 18, 2003. As of September 20, 2007:
http://www.e-publishing.af.mil/shared/media/epubs/AFPAM10-1403.pdf

———, Air Force Operations Planning and Execution, Air Force Instruction 10-401, April 25, 2005.

U.S. Department of Defense, DoD Financial Management Regulation 7000.14R, Vol. 9, Definitions, Summary of Major Changes, May 2005. As of January 4, 2006:
http://www.dod.mil/comptroller/fmr/09/09_definitions.pdf

U.S. General Accounting Office (now U.S. Government Accountability Office), *Military Operations: Contractors Provide Vital Services to Deployed Forces but Are Not Adequately Addressed in DoD Plans*, GAO-03-695, June 2003. As of October 16, 2006:
http://www.gao.gov/new.items/d03695.pdf

U.S. Government Accountability Office, *Global War on Terrorism: DoD Needs to Improve the Reliability of Cost Data and Provide Additional Guidance to Control Costs*, GAO-05-882, September 2005. As of October 16, 2006:
http://www.gao.gov/new.items/d05882.pdf

Worrell, Col Josuelito ("Joe"), briefing, Society of American Military Engineers Joint Senior Noncommissioned Officers Symposium, April 20, 2004.